T0192831

The Earth's Free Oscillations

Oleg V. Petrov

The Earth's Free Oscillations

Formulation and Solution of the Fundamental
Wave Equation of Nature

 Springer

Oleg V. Petrov
Russian Geological Research Institute
Saint-Petersburg, Russia

ISBN 978-3-030-67519-6   ISBN 978-3-030-67517-2   (eBook)
https://doi.org/10.1007/978-3-030-67517-2

This Springer imprint is published by the registered company Springer Nature Switzerland AG
The registered company address is: Gewerbestrasse 11, 6330 Cham, Switzerland

Preface

When studying the relationship between the propagation velocity of various types of bulk and surface seismic waves with radial, spheroidal and torsional oscillations of the Earth having corresponding periods, we are struck by the fundamental problem common to all modern natural science, i.e. the problem of obtaining reference points that allow physical meaning to be attributed to all these discrete oscillatory and continuous wave phenomena that occur in nature. Attempts to unify the relationship of discrete oscillations and the velocity of waves and light propagation occurring in seismology, along with other phenomena associated with gravity and matter, using a visual space-time model in three-dimensional continuous Euclidean space (or, in a more general case, in the four-dimensional space-time continuum), have so far proved unsuccessful. In the present monograph, this relationship is presented through formulations and solutions of the wave equation for the Earth's free oscillations with respect to the special nodal, bifurcational, perspectival and projective reference points within the framework of the three "great geometries" of Euclid, Lobachevsky and Riemann. Using simple and illustrative examples for describing the free oscillations of the Earth and taking into account new visible event horizons related to the velocity of waves and light propagation, we formulated and solved the fundamental wave equation of nature in a form of the three "great theorems": Galilean, Lorentz and Poincaré spatiotemporal transformations.

Saint-Petersburg, Russia   Oleg V. Petrov

Introduction

It is very difficult to come up with new ideas. For this, a completely exceptional imagination is required …

Richard Feynman.
Nobel Laureate
Prize in Physics (1965)

This book addresses the following question: how strongly do people's ideas of the world that surrounds them today remain inextricably linked with the Earth as a solid; moreover, what will happen to these ideas—and geometry as a whole—if the solid Earth starts to oscillate under our feet? These free spheroidal, radial and torsional oscillatory and inextricably interlinked wave displacements of the Earth's bulk and surface occur following each earthquake, causing the planet to "ring like a bell" for many days and weeks at frequencies depending only on the inherent properties of the Earth itself. An exact analogy of these oscillations, referred to in terms of the Earth's free oscillations, is found in a special kind of stringed musical instrument—the monochord—which, as established in Ancient Greece over 2500 years ago, emits musical harmonics that depend only on the length, density and tension of the sounding string. It was this universal music that inspired the French scientist d'Alembert to formulate and solve the one-dimensional wave equation, while the author of the present work offers formulations and solutions of the wave equation for the multi-dimensional free oscillations of the Earth.

Like many other geophysicists who tried to establish a relationship between the propagation velocity of various types of bulk and surface seismic waves with the periods of the radial, spheroidal and torsional eigen oscillations of the Earth corresponding to these seismic waves, the author of the present book ran into the fundamental problem of all modern natural science, i.e. that of obtaining necessary reference points, which make it possible to ascribe a physical meaning of all these discrete oscillatory and continuous natural wave phenomena. As is well-known, all attempts to unify the relationship of discrete oscillations and continuous waves not only in seismology, but also in other phenomena associated with gravity, light and matter in a visual space-time model in three-dimensional continuous Euclidean space (or, in a more general case, in the four-dimensional space-time continuum), have so far proved unsuccessful.

In trying to solve this problem, modern science has borrowed from the empirical features of the propagation of waves and light, becoming convinced that it is only possible to describe them within the framework of the three "great geometries":

– the three-dimensional geometry of the great Greek mathematician Euclid (third century BC) or spherical geometry with constant space curvature;
– the infinite-dimensional geometry of the great Russian mathematician Nikolai Lobachevsky (1826; 1829), having varying negative space curvature;
– the infinite, but still finite-dimensional, geometry of the great German mathematician Bernhard Riemann (1854; 1948) having a changing positive curvature of space.

Using empirical data on the relationship between the propagation velocity of bulk and surface seismic waves and the periods of the radial, spheroidal and torsional free oscillations of the Earth, this new system of measurement will be presented not only with respect to the special nodal and bifurcation perspectives known from Newtonian physics, but also from new perspective and projective reference points associated with the speed of propagation of waves, including light, having the features of three-dimensional Euclidean perspective and projective perception of the world around us as seen through human eyes.

This new fractional measurement system with respect to special nodal and bifurcation—as well as perspective and projective—reference points, combining geometry with physics through the rate of change of the negative and positive curvature of space, as noted by Henri Poincaré, directly follows from the a priori inherent contradictions between three "great geometries", which, like the solutions of d'Alembert wave equation for discrete standing and continuous traveling waves, can in no case avoid being opposed to each other: although there are a priori inherent contradictions between these geometries, it is precisely by opposing them that the new measurement system emerges. These complementary spatiotemporal measurements with respect to special nodal, bifurcation, perspective and projective reference points turn out to be connected with Galilean, Lorentz and Poincaré spatiotemporal transformations of the fundamental laws of nature. This was established taking into account the new visible event horizons that are determined by the speed of light propagation, Schwarzschild gravitational radii and Planck values. Here it seems necessary to remind that the Schwarzschild radius is a radius determined for any object with mass, on whose surface the escape velocity equals the speed of the light propagation, thus making it impossible for the light to escape this object, so for an observer it turns into an invisible projective point in space.

It became apparent that the same indissoluble relationship of complementary space-time measurements with space-time transformations of the fundamental laws of nature with respect to special nodal, bifurcation, perspective and projective reference points forms the basis of the creation of the general and special theories of relativity, quantum mechanics and string theory. It is the same relationship that also forms the starting point for the formulations and solutions of theorems of Henri Poincaré "On two-dimensional and three-dimensional spheres", "On resonances", "On the presence on the surface of an oscillating sphere of at least two pairs of fixed points". Poincaré's theorems are notorious in the history of science for the fact that, although various solutions have been proposed, they turned out to be incorrect upon closer examination. It can be fairly remarked that the three great Euclidian, Lobachevskian and Riemannian geometries have yet to be reliably combined with the unified theory of Galilean, Lorentzian and Poincaréan space-time transformations of the fundamental laws of nature in a unified spatiotemporal system of fractional measurement. In this book, using simple and illustrative examples for describing the free oscillations of a string and the Earth, taking into account the features of the perspective and projective perception of the world around us through human eyes, along with new visible event horizons related to the propagation velocity of waves and light, Schwarzschild radii and Planck values, we continue the work of our great predecessors and present formulations and solutions of the fundamental wave equation of Nature with respect to special nodal and bifurcation, perspective and projective reference points within the framework of the three "great geometries" of Euclid, Lobachevsky and Riemann in the form of three "great theorems", i.e. Galilean, Lorentzian and Poincaréan space-time transformations.

Contents

Chapter 1
Formula of Scientific Discovery: The Fundamental Wave Equation of Nature

Abstract The chapter includes the summary of unified theory of space–time transformations of fundamental laws of nature in the form of three "great theorems" by Galileo, Lorentz and Poincaré. This theory, in light of the initially existing inherent contradictions between three "great geometries" by Euclid, Lobachevsky and Riemann, reflects the dual features of three-dimensional, perspective and projective perception of the world around us through human eyes. Three "great theorems" related to the velocity of light propagation, gravitational radii and Planck's quantities, reveal the essence of unified theory of space–time transformations. Its significance is considered for describing the total energy of oscillatory and wave processes as well as for formulating the fundamental laws of its development in the Universe. In light of unified theory of space–time transformations, these three "great theorems" have the dual form of space–time laws. The last ones unite Newtonian physics and thermodynamics with electromagnetic, gravitational, weak and strong fundamental interactions in regard to special nodal and bifurcation, perspective and projective reference points, which serve as the origin of coordinates of unified metric system.

Keywords Space-time transformations · Theorems by Galileo · Lorentz and Poincaré · Geometries by Euclid · Lobachevsky and Riemann · Radial · Spheroidal and torsional oscillations of the Earth · Reference points

> Reason is the servant of two masters: logic proves, and intuition creates, but only intuition, i.e. comprehension of truth not by way of proof, but by direct intellectual discretion of its content, allows us to make a leap to fundamentally new knowledge.
>
> Henri Poincaré.

Henri Poincaré,
The greatest mathematician of all time.
"The only really valuable thing is intuition…".

Albert Einstein.
Nobel Prize in Physics (1921).

Confirming the indissoluble relationship between the human concept of space/time and the Earth, which, as Albert Einstein rightly noted, is also evidenced by the very name of geometry (from the Ancient Greek: γεωμετρία; geo- "earth", -metron "measurement"), we unite and develop the ideas of the three "great geometries" on the example of the radial, spheroidal and torsional eigenoscillations of the Earth, as well as the bulk and surface seismic waves associated with these oscillations:

- the three-dimensional geometry of the great Greek mathematician Euclid (third century BC) or spherical geometry with constant space curvature;
- the infinite-dimensional geometry of the great Russian mathematician Lobachevsky (1829), having varying negative space curvature;
- the infinite, but still finite-dimensional, geometry of the great German mathematician Riemann (1948), having a changing positive curvature of space.

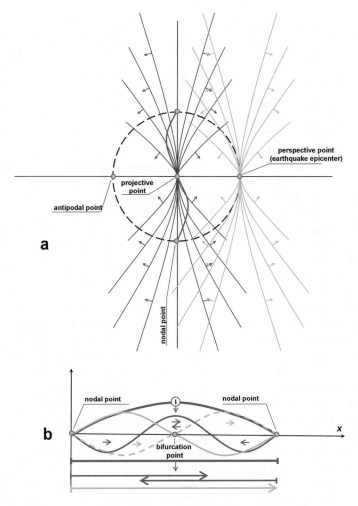

Fig. 1.1 Relationship of the three "great geometries" with discrete nodal and bifurcation, perspective and projective reference points for: the oscillating Earth (**a**); and a vibrating string (**b**)

Within the framework of the three "great geometries", when considering the relationship between, on the one hand, the velocity and propagation time of primary and secondary volumetric seismic waves, as well as surface Love and Rayleigh seismic waves from the epicentre of an earthquake to the antipodal point lying on the opposite surface of the Earth and, on the other hand, the periods of its radial, spheroidal and torsional free oscillations, we were forced to turn to the solution of the famous wave equation of d'Alembert for a vibrating string, often considered by geophysicists as a simplified model of the natural vibrations of the Earth (Fig. 1.1).

These dual solutions of the wave equation for traveling and standing waves, associated with fractal (or fractional) coordinate space and the analytic development of infinites and infinitesimals in integral and differential calculus, are widely known in the history of science as "the vibrating string controversy ". The essence of this debate consists in the attempt to create a spatiotemporal model of the relationship between discrete vibrational and continuous wave phenomena of nature, when the vibrating strings, on the one hand, are considered as standing waves or an infinite set of discrete pendulums, tending to ultimate mass and interconnected by weightless threads or springs, and on the other hand, in the form of continuous traveling waves propagating at a certain speed that is unrelated to the ultimate discrete masses.

Within the framework of contemporary string theory when analysing infinites and infinitesimals in integral and differential calculus during the transition to a new visible event horizon associated with the speed of propagation of waves and light, we first concentrate the entire mass of the vibrating string and sphere at one point, which, within the frame of projective measure, corresponds to their centre of gravity and accounts for the dimensions of their gravitational radius (i.e. a point on the surface of which the escape velocity is equal to the propagation speed of light, becoming an invisible projective point of space for the observer, on which there is a visible event horizon associated with the speed of light propagation). Next, we expand the mass of this point in the form of an endless set of Planck masses tending to the limit, as well as Planck masses on which there is a visible event horizon associated with the speed of light propagation, oscillating on the Planck time scale of discrete pendulums. This relativistic solution of the d'Alembert wave equation for standing waves that takes the form of discrete pendulums interconnected by weightless cords or springs accounts for fundamental gravitational interactions on Planck scales of mass and time through the Poincaré resonances in the form of spatiotemporal Poincaré transformations, which are also lying beyond the visible event horizons. Since the amplitudes of the natural oscillations of invisible discrete Planck masses interconnected by Poincaré resonances will decrease in proportion to the square of the distance from the centre of their concentration, this will also comprise a relativistic or quantum-wave representation for vibrating strings or spheres, as first described in 1667 by Isaac Newton's fundamental law of universal gravitation.

On the other hand, solving the wave equation for a vibrating string relative to a perspective reference point in the form of continuous waves travelling at a speed not exceeding the propagation speed of light, which is not related to the discrete masses, we obtain the famous Albert Einstein equations referred to in the Lorentz space–time transformations

$$x' = \frac{x - \vartheta t}{\sqrt{1 - (\vartheta^2/c^2)}}, y' = y, z' = z, t' = \frac{t - \frac{\vartheta}{c^2}x}{\sqrt{1 - (\vartheta^2/c^2)}}, \qquad (1.1)$$

where c is the propagation speed of light; ϑ is the propagation speed of traveling waves.

These ideas that arose at the beginning of the twentieth century concerning the relationship between the discrete and continuous arising from the relativistic solution of the d'Alembert wave equation for traveling and standing waves in a vibrating string, while in many respects still intuitive, formed the basis for the creation of the general and special theories of relativity, as well as quantum mechanics when moving to new visible event horizons related to the speed of light propagation, Schwarzschild radii and Planck scales. The currently popular string theory concept, which promises to reconcile gravity with quantum theory, arose from the same set of problems. Like those found in everyday life, the strings central to this concept comprise one-dimensional objects, having only the dimension of length. According to this theory, the vibrations of strings moving against the background of spacetime are interpreted as particles.

By the end of the twentieth century, however, it was already becoming clear that string theory refers to only one class of oscillating objects among others not limited to one dimension, of which the eigenoscillations of the Earth comprise a clear example.

In seismology, these oscillating objects consist in the surface-, spheroidal- and torsional free oscillations of the Earth and spatiotemporally-associated Love and Rayleigh surface seismic waves propagating from the earthquake source (or a perspective reference point) along the Earth's surface. The natural vibrations of a two-dimensional sphere and volumetric natural vibrations of the Earth are depicted in Fig. 1.1 in the form of triaxial deformation ellipsoids and radial natural vibrations associated in space–time with the propagation of volumetric primary P- and secondary S-seismic waves, comprising natural vibrations of a three-dimensional sphere.

It turns out that the period of the Earth's radial free oscillations, corresponding to the propagation time of the P- primary and S- secondary bulk seismic waves from an earthquake's epicentre (or perspective reference point) to the antipodal point located on the opposite surface of the Earth, is equal to 20.5 min (Fig. 1.1). When superimposed, the superficial Love and Rayleigh waves propagating along the Earth's surface give torsional modes T having a period of fundamental vibration of 44 min and spheroidal modes S having a period of fundamental vibration of 54 min (Fig. 1.1).

The fundamental radial, spheroidal and torsional oscillations are, in turn, decomposed into an almost infinite number of modes, each characterised by its own oscillation frequency associated with other resonances. Thus, in descriptions of the Earth's natural oscillations, space–time turns out to be related to the propagation velocity of various types of traveling seismic waves and the resonant frequencies of its natural oscillations for standing seismic waves.

At the same time, the innovation of geometric constructions in solving the wave equation for the oscillating Earth consists reflects the need to identify special nodal and bifurcation reference points, as well as additional projective and perspective reference points. The projective reference points necessary for describing the natural oscillations and standing waves of a sphere are forever connected here with the centres of gravity of its discrete masses (Fig. 1.1). At the same time, perspective reference points, necessary for describing the propagation of travelling surface seismic waves that are not associated with discrete masses, always propagate on the surface of a sphere (Fig. 1.1).

In the transition to new visible event horizons related to the propagation speed of waves including light, taking Schwarzschild radii and Planck values into account, we obtain a new spatiotemporal system of measurement in the framework of three "great geometries":

- three-dimensional Euclidean geometry (third century BC) or spherical geometry with constant space curvature;
- infinite-dimensional Lobachevskian or hyperbolic geometry (1826) at speeds not exceeding the propagation speed of light with varying negative space curvature;
- infinite, but still finite-dimensional Riemann geometry (1854) with varying speeds reaching the speed of light propagation, the positive curvature of space in the direction of Schwarzschild gravitational radii and Planck quantities, objects on the surface of which the escape velocity reaches the speed of light propagation and they become black holes and dark matter of the universe invisible to an observer.

From the point of view of topology, it is precisely to perspective and projective measuring systems and Lorentzian and Poincaréan spatiotemporal transformations reflecting fundamental laws of nature in the framework of the three "great geometries" that the formulations and solutions of the famous Poincaré theorems turned out to apply: to the "two-dimensional sphere" as a solution of the wave equation for waves traveling relative to perspective points of reference, whose propagation cannot exceed the speed of light; and to the "three-dimensional sphere" as a solution of the wave equation for standing waves with respect to projective reference points having mass, on the surface of which escape velocity is equal to the speed of light, which becomes invisible to the observer as "black holes" and everywhere-dense dark matter of the Universe (Fig. 1.1).

This new system of metrics, combining geometry with physics in the form of space–time transformations of the fundamental laws of nature formulated by Galileo, Lorentz and Poincaré, directly follows from the inherent contradictions that originally existed between the three "great geometries", which, as it turned out, could by no means avoid being opposed to each other. Combined, they form an inextricably interconnected fractal (or fractional) spatiotemporal measuring system (Figs. 1.1 and 1.2). Thus, the number of lines that can be drawn through a given point parallel to a given line (Fig. 1.2a) is:

- equal to unity in the Euclidean geometry;
- infinite in Lobachevskian geometry;
- zero in Riemannian geometry.

In this sense, the three "great geometries", reflecting the inextricable relationship between discrete and continuous phenomena occurring in nature having the characteristics of three-dimensional Euclidean perspective and projected coordinate space, clearly demonstrate to us the properties of self-similar fractal sets whose existence was first theoretically established by Mandelbrot (1982). Within the framework of the inherent contradictions that initially exist between the three "great geometries",

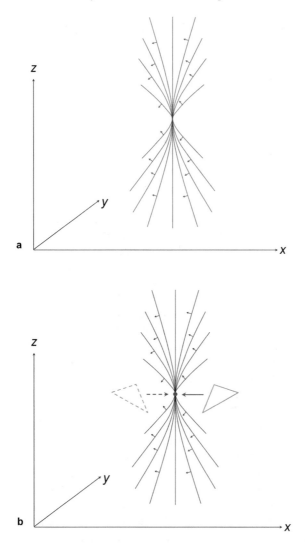

Fig. 1.2 Relationship of the three "great geometries". **a** the sets within the framework of irreducible opposites are similar to the perfect, everywhere dense Cantor set of points which is known in mathematics and which fills the entire segment, although the total mass of its points is zero; **b** the ratio of strong and weak fundamental interactions, including mysterious torsional spins of elementary particles

such sets are similar to the perfect, everywhere-dense Cantor set of points in mathematics that fills the entire Cantor segment, although the total mass of its points is equal to zero (Fig. 1.2a).

Complementing this description with simultaneous oscillations of spaces having changing positive and negative curvatures relative to each other within the framework of the Lobachevskian and Riemannian geometries, taking into account the fact that the sum of the angles of a triangle are:

- equal to two right angles in Euclidean geometry;
- less than two right angles in Lobachevskian geometry with varying negative space curvature;
- greater than two right angles in Riemannian geometry;

we obtain torsional vibrations and surface waves, which, along with the radial energy of the waves reaching the speed of light propagation, determine the structure of electromagnetic and gravitational fields and the nature of strong and weak fundamental interactions, including the mysterious torsional spins of elementary particles (Fig. 1.2c).

Linking the rate of change in the curvature of space with the propagation speed of light, Schwarzschild gravitational radii and Planck quantities, we inevitably encounter the dual discrete-wave nature of matter and the dual nature of standing and traveling waves in the form of natural vibrations of space with respect to special nodal, bifurcation, perspective and projective reference points to arrive at the foundations of quantum mechanics, as well as the general and special theories of relativity, interconnected through space–time transformations of the fundamental laws of nature of Galileo, Lorentz and Poincaré with Newtonian physics.

In this sense, three-dimensional Euclidean, perspective and projective coordinate space and corresponding special nodal, bifurcation, perspective and projective points of origin relative to new event horizons connected to the propagation speed of waves including light are complementary fractal (or fractional) spatiotemporal dimensions, with which a whole network of unexpected relationships between various physical phenomena in the world around us, referred to in dualistic terms, has been associated.

Many fundamental laws of nature here become only an integral part, a perspective or projective fragment of the general picture of the world within the framework of three "great geometries": Lobachevskian geometry with varying negative space curvature, Riemannian geometry with varying positive space curvature and Euclidean geometry or spherical geometry with constant space curvature.

Thus, considering the world around us within the framework of these geometries, with respect to special nodal, bifurcation, perspective and projective reference points and new visible event horizons related to the speed of light propagation, Schwarzschild gravitational radii and Planck values, we reach the frontiers of modern science, where Lorentz contraction and time dilation follow directly from the geometrical structure of space, and Albert Einstein's "observers", "rulers", "clocks", "trains" and "planes" are replaced by external infinite-dimensional Lobachevskian geometry with a varying propagation speed that changes with the speed of light by the negative curvature of space and the internal infinite, but still finite-dimensional Riemannian geometry having a varying speed reaching the propagation speed of light in the direction of Schwarzschild gravitational radii and Planck scales by the positive curvature of space. Here, the disappearance of the curvature of space at all points and for all areas characterises the familiar three-dimensional Euclidean space having a constant curvature.

Following our great predecessors, we present a formula for the scientific discovery of a unified theory of space–time transformations of the fundamental laws of nature of Galileo, Lorentz and Poincaré in the framework of the three "great geometries" of Euclid, Lobachevsky and Riemann, reflecting the features of three-dimensional Euclidean, perspective and projective perception of the world as seen through human eyes and encompassing dimensions from Planck values and Schwarzschild gravitational radii to the propagation speed of light in the Universe in the form of three "great theorems":

Theorem 1 (Galilean space–time transformations) *Within the framework of three "great geometries" of the wave equation for natural vibrations of a string and a sphere with respect to special nodal and bifurcation reference points, this solution implies the same time frame during the transition from one inertial reference system to another.*

From the point of view of topology, this is the solution of the famous Poincaré theorem "on the presence of at least two pairs of fixed points on an oscillating sphere" and the representation of the fractal dimension of oscillating spaces "... as reflections of points, lines and planes in themselves".

Theorem 2 (Lorentz space–time transformations) *This solution is obtained within the framework of three "great geometries" of the wave equation for travelling waves propagating in a perspective measurement system with respect to perspective reference points lying on any spherical surface, when the change in the negative and positive curvature of space cannot exceed the propagation speed of light. From the perspective of the special theory of relativity, these comprise the famous equations of Albert Einstein on the relationship of coordinates in time with the propagation speed of light and on the discrete-wave or quantum nature of light. From a topological point of view, this comprises a solution to the famous Poincaré "two-dimensional sphere" theorem.*

Theorem 3 (Poincaré spatiotemporal transformations) *This solution is obtained in the framework of the three "great geometries" of the wave equation for standing waves in a projective measurement system relative to projective reference points lying inside any spherical surface, on the surface of which the escape velocity is equal to the speed of light propagation, thus comprising observer-invisible "black holes" and the dark matter of the Universe. From the perspective of the general theory of relativity, the surface of such bodies comprises a visible event horizon created by discrete masses that are limiting with respect to the speed of light propagation, whose dimensions correspond to the Schwarzschild radius in the projective observation system, reaching Planck scales and masses in the limit, along with the resonant interrelations between them, which, decreasing inversely in proportion to the square of the distance between them, determine the laws of universal gravitation. From the point of view of topology, this represents the solution of the famous Poincaré theorems "on resonances" and "three-dimensional spheres", as an everywhere-dense set* *"without holes". Such sets within the framework of the inherent contradictions that initially exist between the three "great geometries" are analogous to the perfect, everywhere-dense Cantor set of points in mathematics that fills the entire space, despite the total mass of its points being equal to zero.*

These three "great theorems", based on the inherent contradictions between the three "great geometries", including the dimensions associated with the speed of light propagation, Schwarzschild radii and Planck quantities, reveal the dual essence of formulations and solutions of the fundamental wave equation of Nature in a form of a unified theory of Galilean, Lorentzian and Poincaréan spatiotemporal transformations simultaneously in three-dimensional Euclidean, perspective and projective metrics when describing the total energy of oscillatory and wave processes in the Universe and formulations of the fundamental laws of its development. Within the framework of a unified theory of such spatiotemporal transformations, these fundamental laws of nature take on a dual spatiotemporal form, uniting special-relativistic nodal and bifurcational, as well as perspective and projective reference points, Newtonian physics and thermodynamics with electromagnetic, gravitational, weak and strong fundamental interactions that occur in the "immediate vicinity", either "at the same time", "a little earlier", or "a little later" (Einstein 1936).

This work, initiated at the beginning of the twentieth century by Henri Poincaré and Albert Einstein, to create a unified fractional measurement system and a dual theory of spatiotemporal transformations of the fundamental laws of nature that adequately reflect the reality surrounding us, started to be reflected in the synthetic and synergistic interdisciplinary directions of the development of science at the methodological level from the mid 1950s onwards (Prigogine and Stengers 1986, 2001; Prigogine 2000; Haken 1980, 1985). Today, this dual language of the fundamental laws of nature, confirming the validity and relevance of the statements of Poincaré and Einstein concerning the role of scientific intuition in "…comprehending the truth not by proving it, but by direct intellectual discernment of its content…", can be easily clarified using a vibrating string as an example and considering the Earth's free oscillations in terms of a simple school globe as part of a high school programme.

References

Einstein A (1936) Physik und Realitat. J Frankl Inst 221: 313–347. (Russian Translation: Einstein A (1967) Physics and reality. Einstein A (1967) Selection of scientific works in 4 volumes. M. Nauka, vol 4, p 200–227)
Haken H (1980) Synergetics. M.: Mir, p 404. (in Russ.)
Haken H (1985) Synergetics. Hierarchies of instabilities in self-organising systems and devices. M.: Mir, p 419. (in Russ.)
Lobachevsky NN (1829) Kazan bulletin. Part XXVI, books V and VI, May–June, (in Russ.)
Mandelbrot BB (1982) The fractal geometry of nature. WH Freman, San Francisco, p 460
Prigogine IR (2000) The end of certainty. Time, chaos and new laws of the nature. Scientific publishing centre «Regular and chaotic dynamics», Izhevsk, p 208. (in Russ.)

Prigogine IR, Stengers I (1986) Order out of chaos: man's new dialogue with nature. In: Arshinov VI, Klimontovich YuL, Sachkov YuV. M: Progress, p 432. (in Russ.)

Prigogine IR, Stengers I (2001) Time, chaos, quantum. To the solution of the time paradox—M: Editorial URSS, p 240. (in Russ.)

Riemann B (1948) Works. M–L: GITTL, p 291. (in Russ.)

Chapter 2
Description of Scientific Discovery: The Fundamental Wave Equation of Nature

Abstract In the chapter, the prerequisites and basic concepts of unified theory of space–time transformations of fundamental laws of nature in the form of three "great theorems" by Galileo, Lorentz and Poincaré are considered in detail in the context of three "great geometries" by Euclid, Lobachevsky and Riemann using simple and obvious cases of description of free oscillations of string and the Earth. The author provides formulations and solutions of wave equation for free oscillations of string and radial, spheroidal, torsional free oscillations of the Earth as well. Considering these formulations and solutions from the point of view of new visible event horizons associated with the velocity of light propagation, gravitational radii and Planck's quantities, the author comes up with the formulation of unified theory of space–time transformations of fundamental laws of nature. From the point of view of exact sciences—mathematics, geometry and physics, the unified theory of space–time transformations of fundamental laws of nature in the form of three theorems by Galileo, Lorentz and Poincaré is supposed to be related to further development of infinitely large and infinitesimal in integral and differential calculus in the transition to new visible event horizons, determined by velocity of waves and light propagation, gravitational radii and Planck's quantities. The chapter also includes the consideration of applied aspects of unified theory of space–time transforms of fundamental laws of nature, in particular, the relation of the Earth's free oscillations to standing internal gravitational waves that determine the fractal hierarchy of dissipative structures of the Earth, which is associated with many aspects of its geological structure and evolution.

Keywords Space-time transformations · Theorems by Galileo · Lorentz and Poincaré · Geometries by Euclid · Lobachevsky and Riemann · Free oscillations of string · Radial · Spheroidal and torsional oscillations of the Earth · Standing internal gravitational waves

> Examining the world around us in three-dimensional space, we are each time surprised to find that a significant part of the concrete world surrounding us escapes through the mesh of the scientific net.
>
> Alfred North Whitehead.

Alfred North Whitehead.
Developer of the foundations of logicism and type theory.

For millennia, mankind has been verifying the conclusions of Euclidean geometry, applying them in its practical activities and making sure that they correctly reflect the spatial and temporal relations of the real world. However, with Isaac Newton's discovery that the fundamental laws of nature, geometry and physics were inextricably interlinked, it also turned out that the relationship of vibrational and wave phenomena of nature cannot be fully described solely in three-dimensional continuous space (or, in the more general case, in a four-dimensional space–time continuum).

Following the statement and solution of the wave equation in 1752 by the outstanding French mathematician Jean le Rond d'Alembert, this interlinking became the basis of a scientific discussion that unfolded in the middle of the eighteenth century. The ensuing discussion, widely known in science as the "vibrating string controversy", also laid the theoretical foundations for contemporary "string theory".

From a mathematical perspective, the essence of this controversy is connected with the analysis of infinitesimals and infinites in differential and integral calculus, concluding in the attempt to create a model of the relationship between oscillatory and wave phenomena of nature. Here, vibrating strings, on the one hand, are considered as standing waves or an infinite set of discrete tending-to-ultimate-mass pendulums, interconnected by weightless threads or springs, and on another, in the form of continuous traveling waves propagating at a certain speed, unrelated to the ultimate discrete masses.

From the standpoint of geometry and physics, the study of the relationship between oscillatory and wave phenomena of nature on the basis of experimental data on the propagation of sound and light waves led to the creation in 1826 by Nikolai Lobachevsky of external infinite-dimensional geometry with negative space curvature changing with a speed not exceeding that of light propagation, as well as, in 1854, to the creation by Bernhard Riemann of an internally infinite, but still finite-dimensional geometry with varying speed reaching the speed of light propagation and positive curvature of space in the direction of discrete objects, the dimension of which turned out to be connected with Schwarzschild radii and Planck scales. On the surface of these discrete objects, from which the escape velocity reaches the propagation speed of light, they become invisible to the observer as "black holes" and the dark matter of the Universe. Of course, awareness of the physical nature of these new visible event horizons associated with the propagation speed of light, Schwarzschild gravitational radii and Planck magnitudes came into geometry much later, as well as an understanding of the inextricable relationship of these three "great geometries", which in no case, like the famous the d'Alembert wave equation for standing and traveling waves, could avoid being opposed to each other.

At the beginning of the twentieth century, in an attempt to recognise this new dimension of the world around us, including the propagation speed of light, Schwarzschild radii and Planck scales, Albert Einstein formulated the general and special theory of relativity along with the foundations of quantum mechanics.

In his attempt to expand the dimensions of space and time, Henri Poincaré created the new scientific discipline of topology, which considers the peculiarities of the three-dimensional, perspective and projective mapping of the world around us onto the human retina. This led to the formulation of his famous theorems "On two-dimensional and three-dimensional spheres", "On resonances" and "On the presence of at least two pairs of fixed points on the surface of an oscillating sphere". Poincaré also developed a new methodology for studying the spatiotemporal relationship of oscillatory and wave phenomena of nature from the standpoint of three-dimensional, perspective and projective metrics, which led in the direction of a solution to these theorems.

Developing this new methodology for studying the space around us with a constant and variable speed not exceeding the speed of light propagation, including negative and positive curvature of space from the standpoint of three-dimensional, perspective and projective metrics, we are faced with a difficult task: how to bring the essence of our scientific discovery to the wider scientific community. Indeed, in some cases, this is hidden in the depths of a complex and highly abstract mathematical apparatus. Moreover, although this abstract mathematical apparatus is already to a large extent in use today, it does not yet fully reflect the new features of the perspective and projective perception of the real world through human eyes. It is here that graphical— or topological—methods, used to describe the natural vibrations of the string and the Earth and the propagation of bulk and surface waves associated with these vibrations, come to our aid.

In principle, these methods resemble the well-developed digital technologies based on highly-accurate proofs developed by Hassler Whitney and others within the framework of Lobachevskian, Riemannian and Poincaréan geometry. The main difference here consists in the use of various graphical constructions to reduce the calculations, which take into account perspective and projective features of the perception of the world around us as seen through human eyes. Moreover, although at a qualitative level leading to the same results as numerical methods, the topological method is relatively simple to use due to its highly visual character. This approach becomes especially useful in the initial investigation of a problem, since the qualitative nature of the solution can in some cases be determined relatively quickly.

These graphical measurement algorithms, confirming the inextricable relationship of geometry with physics using simple and illustrative examples of the free oscillations of a string and the Earth, are the logical conclusion of the formulations of the fundamental laws of nature in the framework of three "great geometries":

- three-dimensional Euclidean geometry (third century BC) or spherical geometry with constant space curvature;
- infinite-dimensional Lobachevskian or hyperbolic geometry (1826) at speeds not exceeding the propagation speed of light with varying negative space curvature;
- infinite, but still finite-dimensional Riemannian geometry (1854) with varying speeds reaching the speed of light propagation, the positive curvature of space in the direction of Schwarzschild radii and Planck quantities, objects on the surface of which the escape velocity reaches the speed of light propagation and they become black holes and dark matter of the universe invisible to an observer.

Despite its apparent simplicity, this essentially novel approach to describing physical reality involving three-dimensional, perspective and projective methods of spatial constructions, widely known in architecture and painting since the Renaissance, provides the basis for the algorithms of contemporary mathematics—more abstract, but also more powerful than those of classical Newtonian or quantum mechanics, dynamics and thermodynamics, or even the general and special theories of relativity.

2.1 Formulation and Solution of the Wave Equation for Free Oscillations of a String

The history of science from Newton to Einstein attests to the fact that many mysterious natural phenomena "… stem from an inability to accurately describe physical phenomena in three-dimensional continuous space or (in a more general case) in the four-dimensional space–time continuum. Although it is possible that someday, on the basis of other ideas, we will be able to penetrate deeper into the meaning of these great guiding principles of physics beyond the Standard Model (BSM), today they remain completely obscure."
Louis de Broglie (1965).

Louis de Broglie.
Laureate Nobel Prize in Physics (1929).

In the history of the natural sciences, awareness of the inextricable relationship of geometry with physics is reflected not only in the creation of the theory of relativity and quantum mechanics, but, above all, in the scientific "controversy of the vibrating string".

As far back as ancient Greece, a special one-stringed musical instrument known as the monochord was used to study string vibrations, including the mathematical relations so illustrated, which were widely spread thanks to the school of Pythagoras (sixth century BC). Even then it was known that, along with the main tone, the string emits additional sounds called partial tones, or overtones.

Indeed, it is by turning to musical language that we can easily envisage the vibrations that occur when plucking a violin string. Plucking the string creates a musical tone consisting of the fundamental tone and the overtones—or higher harmonics— of the string. It was precisely as a result of the study of these physical phenomena, referred to as free- or eigenoscillations, that the mathematical theory of the Fourier series arose.

At first glance, it seems that the movements of the string can easily be explained either by waves traveling back and forth along the string after being reflected at its ends, or in the form of standing vibrations of the string perpendicular to its length (Fig. 2.1). However, the equal validity of these two interpretations, as well as the solution of their corresponding equations, is far from obvious. In the case of a traveling wave, there are not only periodic discrete vibrations occurring at each point of the body along which the wave propagates, but also continuous transmission of the vibrational state through the body of the string.

Following the formulation and solution by the eminent French mathematician and philosopher d'Alembert of the so-called wave equation, this question became the basis of the scientific discussion that unfolded in natural science during the middle of the eighteenth century. About two centuries later, the discrete oscillatory and continuous wave properties of matter would again show themselves in a much more mysterious form in the field of light phenomena than in the material phenomena of the macrocosm. It turned out that the discrete oscillations described by differential

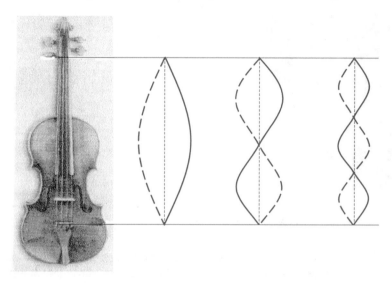

Fig. 2.1 Modes of free oscillations of a string. The string moves from the position shown by the solid line and vice versa. The fixed points are known as nodes (Hawking 2007)

equations are far from fully capturing the entire wealth of oscillatory processes occurring in the real world. At the same time, distributed processes described by partial differential equations or integral equations proved to be adequate. Since these wave equations represent collective oscillatory movements in distributed systems, they do not have point analogues. Thus, at the level of the invisible microcosm, the formidable problem of the dual nature of oscillations and waves, which became known as wave-particle duality, was posed (Fig. 2.2).

For almost three hundred years, the discussion of the dual nature of the discrete vibrational and continuous wave properties of matter has repeatedly shaken the edifice of the natural sciences. Thus, it is no accident that the fundamental provisions of this discussion in various aspects with respect to continuous media and subtle physical continua have been addressed by such luminaries as Leonard Euler, Jean d'Alembert, Joseph-Louis Lagrange, Henri Poincaré, Louis de Broglie, Erwin Schrödinger, Max Planck, Max Bourne, Albert Einstein, Paul Dirac, John von Neumann, Georg Cantor, William Hamilton, Aleksandr Lyapunov, Andrey Kolmogorov, Vladimir Arnold, Jürgen Moser, Leonid Mandelstam, Aleksandr Andronov and Ilya Prigogine. Today, the answer to this question remains just as relevant as three hundred years ago, since, from the point of view of the theory of relativity, the vibrating string has again gained wide prominence in contemporary string theory as a fundamental one-dimensional object that replaces the idea of elementary particles that do not have an internal structure.

Why has this discussion so far failed to lead to a conclusion? Why, when describing the free oscillations of the Earth and objects of the invisible microcosm, did the vibrating string regain its position of relevance in contemporary science? We will

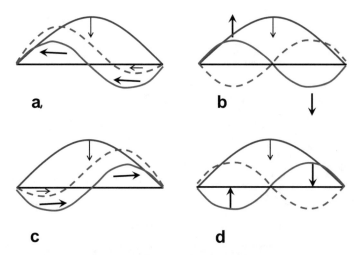

Fig. 2.2 Oscillations of a string, represented as running-, from right to left (**a**) and from left to right (**b**), or standing–(**c, d**) waves

attempt to answer this question from the perspective of topology—the science of the structure of space and its dimension, which takes into account the features of the perspective and projective reflection of the world around us on the retina of the human eye.

Let's consider a string fixed at two nodal points (Fig. 2.3) and try to describe its movement.

If we introduce three independent observers x, y, z in this reference frame, as shown in Fig. 2.3, then from the position of topology we get the full algorithm of free oscillations of an x, y, z three-dimensional string in x, y, z three-dimensional Euclidean space. Obviously, in this case, the oscillations are split into transverse and longitudinal ones.

The transverse oscillations are seen by observer y relative to (x, z) projective plane and the longitudinal oscillations are observed by observer z, which describes the oscillations of the (x, y, z) three-dimensional string's bulk relative to the projective line formed by the intersection of (x, z) and (x, y) projective planes (Fig. 2.3).

Longitudinal or primary bulk P-waves displace the particles of the vibrating string in the horizontal direction and create tension and compression zones (Figs. 2.3. and 2.4.).

All these longitudinal and transverse waves are described by the same wave equation:

$$\frac{1}{c^2}\frac{\partial^2 u}{\partial t^2} = \frac{\partial^2 u}{\partial x^2} + \frac{\partial^2 u}{\partial y^2} + \frac{\partial^2 u}{\partial z^2}, \tag{2.1}$$

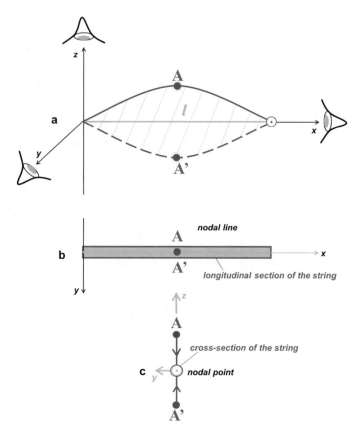

Fig. 2.3 Topology of the normal mode of a string's free oscillations from the point of view of independent projective observers x, y, z, who simultaneously see its movement on the screens of two dimensions relative to the nodal plane (**a**), nodal line (**b**) and nodal point (**c**). The string moves from the position indicated by solid red line and point "A" to the position indicated by dashed red line and point "A", and back: observer y will see plane oscillations and waves (**a**); observer z will see bulk oscillations and waves (**b**) while observer x will observe discrete oscillations and spherically symmetric waves (**c**)

where c has a dimension of velocity and depends on the density of the material string ρ; however, the propagation velocity of longitudinal waves $c_P = \sqrt{E/\rho}$ is determined by the elasticity of the substance (E is Young's modulus), and of the transverse waves $c_S = \sqrt{T/\rho}$ is determined by the tension of the string.

According to Hooke's law, deformation ε is associated with T tension force in the string by $\varepsilon = $ T/E or T $= \varepsilon E$, giving:

$$\frac{\rho}{E}\frac{\partial^2 u}{\partial t^2} = \frac{\partial^2 u}{\partial x^2} + \frac{\partial^2 u}{\partial y^2} + \frac{\partial^2 u}{\partial z^2} \text{ for longitudinal waves and}$$

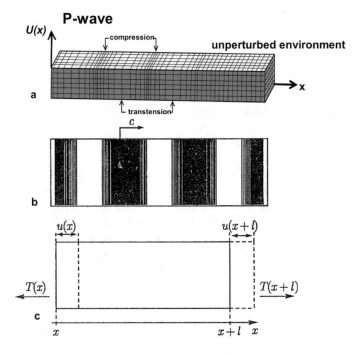

Fig. 2.4 Longitudinal compression-extension waves in a string—(**a**); **b**—dashed lines show the new position of the considered bulk element; **c**—the determination of the displacement vector of the particle substance: T is the tension force in the string, $u(x)$ and $u(x + 1)$ are the functions indicating the distance by which the points with x and $x + 1$ coordinates, respectively, have displaced

$$\frac{\rho}{E \cdot \varepsilon} \frac{\partial^2 u}{\partial t^2} = \frac{\partial^2 u}{\partial x^2} + \frac{\partial^2 u}{\partial y^2} + \frac{\partial^2 u}{\partial z^2} \text{ for transverse waves.}$$

Moreover, since the velocity of longitudinal waves is always greater than transverse, they are called primary.

Thus, from the point of view of projective geometry, we get the opportunity to simultaneously describe the string's oscillations relative to the (x, z) projective plane in the form of plane waves, by the projective line, which is the intersection of two (x, z) and (x, y) projective planes in the form of bulk longitudinal waves and a nodal point, which can also be viewed from the position of an observer x as an infinite number of points that belong to the nodal line, blocking each other, which will inevitably lead us to the Newtonian model of the atom structure and the Newtonian interpretation of quantum mechanics. This was exactly the manner in which many key ideas of modern geometry and physics were developed—including the singular points of Newtonian dynamical systems—and several fundamental laws formulated. However, even this description of a vibrating string turned out to be far from complete. The outstanding French mathematician d'Alembert in the eighteenth century not only wrote out the first partial derivatives equation in mathematical physics, known as the wave equation.

$$u_{tt} - c^2 u_{xx} = 0, \quad c^2 = F_0 l_0 / m \tag{2.2}$$

where u_{xx} is the partial derivative with respect to x, u_{tt} is the partial derivative with respect to time, F_0 is the string tension force, l_0 is the string's length, m is the string's mass,

but also obtained its general solution.

$$u = f(x - ct) + g(x + ct) \tag{2.3}$$

This is a very interesting formula, which is convenient to use when the disturbance has not yet reached the ends of the string. Indeed, if we disturb the balance of the string, we will find, along with the vertical, the horizontal oscillatory movements, which propagate to the right and left relative to its nodal points (Fig. 2.5). With the propagation of horizontal oscillations running towards each other, standing waves arise that correspond to the second overtone of the vibrating string. They are shown in Fig. 2.6. in red. Further propagation of horizontal oscillations of the string is possible only to the right or left. Here, the string's horizontal and vertical oscillations branch. In Fig. 2.6. these string oscillations in the form of the first overtone are shown in green.

From the standpoint of modern mathematics, when solving the equation for the transverse and longitudinal oscillations (waves) of a string, it is to be expected that we will encounter a bifurcation of these solutions. This concept, which is typical of so-called nonlinear equations, will be explained with a classic example.

Let's imagine, instead of the central part of the string, a column on which load P acts from above (Fig. 2.7a and b). We will increase the load and observe what happens. At first, the column will be shortened and thickened, but its centre line will remain straight. However, at a certain critical value P_C, the picture will qualitatively change—the column will lose its rectilinear shape and will bend to the right or left.

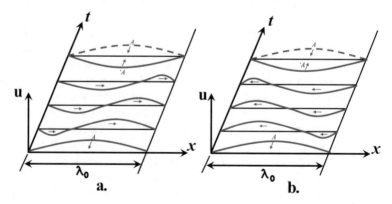

Fig. 2.5 The general solution of the d'Alembert wave equation $u = f(x–ct) + g(x + ct)$ in the form of waves traveling to the right (**a**) or left (**b**). λ_0 corresponds to the eigenvalue (or period) of the natural mode

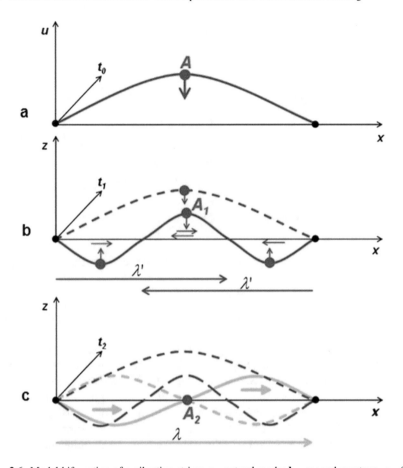

Fig. 2.6 Modal bifurcation of a vibrating string: **a**—natural mode; **b**—second overtone, **c**—first overtone. At different time intervals (t_0, t_1, t_2), arrows show different displacement directions for standing and traveling plane waves of a vibrating string

At $P < P_C$, the column has a single equilibrium form. At $P > P_C$ there are three possible forms: a rectilinear form that has become unstable, and two stable forms—one corresponding to a deflection to the right, while the other deflects to the left. If we draw the dependence of u deviation of the column's axis on P direct load, then the resultant picture will be as shown in Fig. 2.7.

At $P = P_C$, the number of equilibrium states and their stability has changed. The alteration in the number and stability of solutions is referred to in terms of branching—or bifurcation—of the solution. Many outstanding mathematicians such as Leonhard Euler, Jacob and Johann Bernoulli and Joseph-Louis Lagrange were engaged in the problem of the loss of column's stability. One of the first to introduce the term "bifurcation" was Carl Gustav Jacob Jacobi back in 1834. However, the full significance of the theory of bifurcation was realised only at the end of the nineteenth

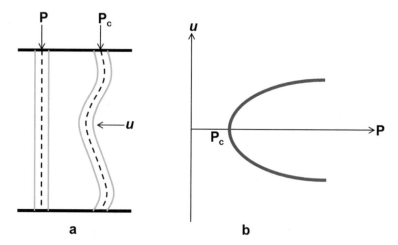

Fig. 2.7 Bending of a column under load (**a**). Dependence of the deviation from the magnitude of the load (**b**)

century by Henri Poincaré, who founded the new direction in the studies of space structure and dimension known as topology.

In connection with the ongoing vibrating string controversy, topology led to the discovery of new system for describing bifurcation, which was necessary in order to account for the dual nature of oscillations (Fig. 2.8). It turned out that as oscillating motions develop as a result of bifurcation, the vibrations "split" into standing and traveling waves. Further, as can be seen in Fig. 2.9, in a vibrating string simultaneously with the first bifurcation and the appearance of the first overtone for traveling waves, the second overtone "splits" relative to the fixed nodal points into standing

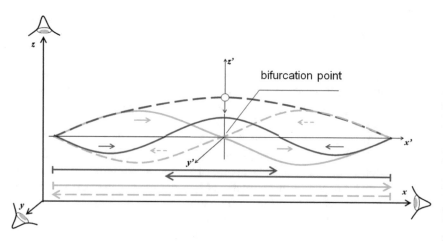

Fig. 2.8 Bifurcation point in a vibrating string

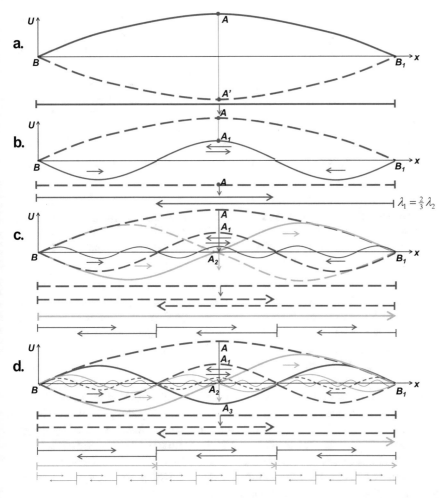

Fig. 2.9 Modal bifurcation of a vibrating string: **a**—natural mode; **b**—natural mode and second overtone; **c**—natural mode, and first, second and third overtones; **d**—natural mode, and first, second, third, fourth and fifth overtones. Green colour—bifurcation modes

waves associated with an overtone of higher order. At the next step, as a result of a new bifurcation and the appearance of traveling and standing waves related to overtones of an even higher order, the vibrational and wave motions are even more complicated (Fig. 2.9).

It is impossible to accurately predict how a string will behave relative to a new coordinate system, which consists in a bifurcation point that corresponds to the centre of gravity of a vibrating string.

If we now concentrate the entire m mass of the vibrating string in its centre of gravity in the form of a weight suspended on a spring, then, as the oscillations are deconvoluted into modes, we obtain the m mass dispersion along its entire length

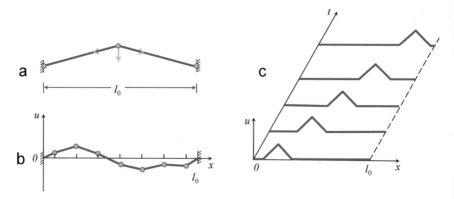

Fig. 2.10 Modelling of string vibrations, on the one hand, in the form of an infinite set of oscillating pendulums tending to the limit mass, connected by weightless threads or springs (**a**, **b**), and, on the other hand, in the form of continuous traveling waves propagating at a certain speed ($\vec{\vartheta}$), not related to limiting discrete masses (**c**)

(Figs. 2.6 and 2.10a, b). Despite the fact that Johann and Daniel Bernoulli, who are also known as outstanding hydrodynamicists, argued in the same way back in the 30 s of the eighteenth century, these achievements of mathematical physics made it possible to obtain the dual formulations of the wave equation for a vibrating string from the simplest laws of nature determining the relationship between vibrational and wave phenomena. The essence of these dual formulations is an attempt to create a direct model of the relationship between vibrational and wave phenomena, when the vibrating strings are considered, on the one hand, as standing waves or an infinite set of discrete pendulums tending to the limiting masses, interconnected by weightless threads or springs, and on the other, in the form of continuous traveling waves propagating at a certain speed, unrelated to the ultimate discrete masses. These dual formulations of the wave equation in relation to standing oscillating and traveling wave phenomena, which turned out to be connected with the development of the analysis of infinitesimal values in differential calculus and infinitely large values in integral calculus, are widely known in the history of science as the "vibrating string controversy"; this controversy persists to the present day in the form of so-called string theory.

The dual formulations of the wave equation and further analysis of infinitesimal and infinitely large values in differential and integral calculus on the example of a vibrating string, when the latter is considered, on the one hand, as an infinite set of oscillating discrete pendulums tending to the limit mass, interconnected by weightless threads or springs and, on the other hand, in the form of continuous traveling waves propagating at a certain speed and not related to limiting discrete masses, established the possibility of special perspective and projective reference points for a vibrating string in addition to special nodal and bifurcation reference points.

The concept of a perspective reference point here is associated with the propagation of traveling waves that move at $\vec{\vartheta}$ speed in their own reference frame x' *relative* to an observer (Fig. 2.10c).

Let's suppose that the wave profile does not change over time, but only its position in space changes (Fig. 2.10c). This means that at any time, the wave profile in the associated reference frame has the form shown in Fig. 2.10c.

Let the wave move at a constant speed along the x axis. Then the coordinates in the fixed (x) and moving (x') systems are interconnected by the ratio:

$$x\prime = x - \vartheta t, \quad y\prime = y, \quad z\prime = z, \quad t\prime = t \tag{2.4}$$

As is known, this system of equations is referred to in terms of Galilean space–time transformations, which imply the same time in all reference systems ("absolute time"). Along with similar intuitively obvious ideas of the symmetry of space and the principle of superposition, which assert the equivalence of the interaction of many bodies in a vibrating string in a short period of time, the Galilean transformations were considered sufficient grounds for formulating the wave equation in the framework of Newtonian mechanics of three-dimensional Euclidean mensuration, based on locality associated with a small field of clear vision of a person.

However, during the transition to new visible event horizons related to the speed of wave and light propagation, the space–time transformations of coordinate systems for a perspective reference point turned out to be determined by well-known equation formulated by Albert Einstein:

$$x\prime = \frac{x - \vartheta t}{\sqrt{1 - \left(\vartheta^2/c^2\right)}}, \quad y\prime = y, \quad z\prime = z, \quad t\prime = \frac{t - \frac{\vartheta}{c^2}x}{\sqrt{1 - \left(\vartheta^2/c^2\right)}} \tag{2.5}$$

where c is the speed of light propagation, ϑ is the speed of the perspective reference point (or traveling waves). This system of equations is referred to as the Lorentz space–time transformations. Moreover, if instead of the law of light propagation we "tacitly proceed" from the ideas of the "old" Newtonian mechanics about the absolute nature of time and the extent of space, then instead of Einstein's equations, we would obtain.

$$x\prime = x - \vartheta t, \quad y\prime = y, \quad z\prime = z, \quad t\prime = t. \tag{2.6}$$

Galilean space–time transformations, which are derived from the Lorentz space–time transformations, would provide us with the principles of simultaneity and symmetry if in the latter c speed of light is set equal to an infinitely large value.

The new visible event horizon associated with the speed of light propagation formed the basis of the Lorentz space–time transformations, which, in turn, formed the basis of the special and general theory of relativity, quantum mechanics and contemporary string theory.

This would not be surprising if it wasn't the case that the same visible event horizon associated with the speed of light propagation determines in the limit the second form of the relativistic solution of the d'Alembert wave equation for gravitational standing waves in a vibrating string. In this case, the entire mass of the vibrating string is first concentrated in its centre of gravity at the projective origin in such a way that when moving to new visible event horizons related to the speed of light propagation, the projective origin becomes a physical body with mass, on the surface of which, from the standpoint of classical Newtonian physics, the escape velocity is equal to the speed of light propagation. Thus, it becomes a "black hole", invisible to the observer.

The importance of such objects in the perception of the world around us was first noticed by John Michell.[1] In a letter sent to the Royal Society dated November 27, 1783, he attempted to reconcile Newtonian celestial mechanics and corpuscular optics. In this letter, he also described the concept of a massive body, whose gravitational attraction was so great that the speed necessary to overcome this attraction (escape velocity) would be equal to or greater than the speed of light. He even included a calculation from which it followed that, for a body with a radius of 500 solar radii and with a density no less than that of the Sun, the escape velocity on its surface will be equal to the speed of light (Ellis 1999). Thus, since no light can escape such a body, it will be invisible to an observer (Levin 2005). Michell suggested that, while many such objects inaccessible to direct observation may exist in space, their existence can be inferred by their gravitational effect on other bodies orbiting them.

In developing these ideas in 1916, almost simultaneously with the creation of the general and special theory of relativity by Albert Einstein, the German scientist Karl Schwarzschild introduced the concept of "gravitational radius" into scientific usage. From the point of view of the general theory of relativity and projective mensuration, when moving to a new visible event horizon related to the speed of light propagation, the gravitational (or Schwarzschild) radius is a characteristic radius determined for any physical body having a mass on the surface of which there is a visible event horizon created by this mass in relation to the propagation speed of light.

The gravitational radius in this sense, proportional to the mass M of the body, is equal to $r_g = 2GM/c^2$ where G is the gravitational constant, and c is the speed of light in vacuum. The gravitational radius of ordinary astrophysical objects is negligible compared to their actual sizes: for example, for the Earth it is $r_g = 0.887$ cm, for the Sun $r_g = 2.95$ km. When moving to the Planck scale of lengths of $l_p = \sqrt{G/c^2} \cdot \hbar \approx 10^{-35}$ m, it is convenient to record it in the form $r_g = 2(G/c^3)M_c$, from which the Planck mass $M_c = 2.176 \cdot 10^{-8\,\mathrm{kg}}$ follows. It is these masses having a maximum possible accuracy to a dimensionless coefficient close to 1, in the mass spectrum of elementary particles having Planck dimensions, as well as remaining "black holes" invisible to the observer, the dark matter of the Universe is formed in our proposed relativistic solutions of the d'Alembert wave equation. The assumption of the existence of these particles, which have been called maximons, planxons or gravitons in modern literature, was made back in 1966 by Markov (2000).

---

[1] Oxford Dictionary of National Biography, 2004.

From the point of view of projective mensuration, it is with respect to the visible event horizons on the surface of such bodies created by masses limited with respect to size, that resonant interactions at the level of Planck time, Planck dimensions and Planck masses come to determine the laws of gravity and the existence of dark matter in the Universe.

Thus, solving the d'Alembert wave equation for a vibrating string in the framework of relativistic physics, we first concentrate the entire mass of the vibrating string at one point, corresponding to its centre of gravity and the dimensions of its gravitational radius, i.e. it becomes an invisible projective point of space for the observer on which there is a visible event horizon associated with the speed of light propagation. Next, we expand the mass of the string in the form of an infinite set of oscillating discrete pendulums tending to the maximum Planck sizes and Planck masses, providing fundamental gravitational interactions through Poincaré resonances, which also lie beyond the boundaries of visible event horizons. Since the amplitudes of natural vibrations of invisible discrete Planck masses will decrease in proportion to the square of the distance from the centre of their concentration, this will be the relativistic or quantum-wave representation of Newton's well-known law describing fundamental gravitational interactions.

Unlike most other Planck quantities, the Planck mass is close to scales that are familiar to the human eye. Thus, a flea has a mass of 4000–5000 Mp, while the sizes of these objects correspond to Planck values of the order of 10^{-33} cm. In this sense, solving the well-known d'Alembert wave equation with respect to the string free oscillations in the form of an infinite set of discrete pendulums tending to Planck masses, scales and times of resonant interaction between the "weightless threads connecting them", we develop infinitesimal values in differential and infinitely large values in integral calculus with respect to new event horizons related to the propagation speed of light to create a visual model of fundamental gravitational interactions. These projective, spatiotemporal transformations of the world around us, lying beyond the boundaries of visible event horizons and relative to the invisible projective reference points, are referred to as Poincaré spatiotemporal transformations.

Thus, using simple and illustrative examples for describing the free vibrations of a string with respect to special nodal and bifurcation perspective and projective reference points during the transition to new visible event horizons related to the speed of wave (including light) propagation, Schwarzschild radii and Planck values, we represent spatiotemporal transformations of the fundamental laws of nature formulated by Galileo, Lorentz and Poincaré in the form of a simple and intuitive spatiotemporal model that reflects the features of the three-dimensional Euclidean space, as well as perspective and projective perception of the world as seen through human eyes. These new formulations and solutions of the well-known d'Alembert wave equation for a vibrating string with respect to special nodal, bifurcation, perspective and projective reference points during the transition to new visible event horizons associated with the speed of light propagation, Schwarzschild radii and Planck values, not only confirm the courageous theoretical constructions of our predecessors, but also their fundamental ideas about the unity of the laws of nature and the possibility of their speculative comprehension.

2.2   Formulation and Solution of the Wave Equation for the Earth's Free Oscillations

New horizons open at selected moments in the history of science.
Ilya Prigogine.

Ilya Prigogine.
Nobel Prize winner in chemistry (1977).

Since the vibrating string is often considered by geophysicists as a simplified model of the Earth's free oscillations, the time has come to make a fundamental step forward in creating a new geometry of nature based on formulations and solutions of the wave equation for the oscillating Earth, reflecting both the features of three-dimensional Euclidean perspective and projective perception of the world around us through human eyes, as well as new visible event horizons associated with the velocity of wave and light propagation.

The history of the study of the Earth's free oscillations is connected with the names of the famous scientists, such as Siméon Poisson, Horace Lamb, James Jeans, Lord Rayleigh and Augustus Love, who first classified the free oscillations of the elastic sphere and then generalised the equations of elasticity theory for gravitating bodies, necessary in considering oscillations of planetary dimension bodies.

The radial oscillations of the sphere were already being studied in 1828 by Poisson, who, together with Augustin-Louis Cauchy, laid the foundations of the linear elastic theory (Fig. 2.11). In 1882, in one of his fundamental theoretical works on seismology, the English mathematician Horace Lamb proved the existence of two independent types of free oscillations for an elastic spherical shell additional to the known radial type (Lamb 1882). The first type includes the so-called seismological S-modes or spheroidal oscillations, which assume the displacements of the sphere particles to have both radial and horizontal components (Fig. 2.12).

The second type includes torsional oscillations (T-modes). During oscillations of this type, the displacements, which are directed tangentially to the surface of the full sphere, have no radial component. Since spheroidal and torsional oscillations are

S U R F A C E

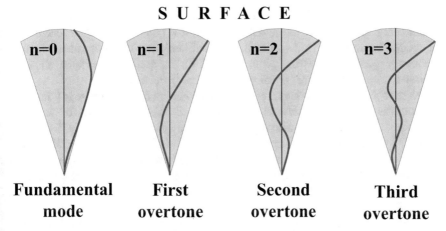

Fundamental mode   **First overtone**   **Second overtone**   **Third overtone**

Fig. 2.11 Radial modes of free oscillations in a full sphere. The curves demonstrate the in-depth displacement for a uniform full sphere

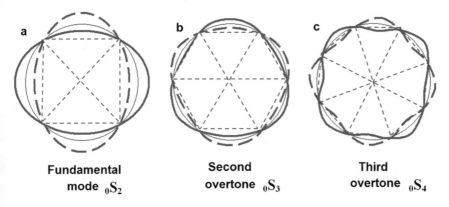

Fundamental mode $_0S_2$ **Second overtone $_0S_3$**   **Third overtone $_0S_4$**

Fig. 2.12 The full sphere surface oscillations in a profile section. The intersection points of the red solid and dashed lines lie on the nodal lines (blue dotted line) at the surface of the full sphere

usually found together, the displacements at each point on the surface are represented by a combination of both types of oscillation.

The easily comprehensible classification of the T- and S-modes is based on the image of the motion scheme represented as a grid of nodal surfaces, i.e. places of the absent motion. To this end, imagine for a while that there is no liquid Earth's core and the Earth is a simple irrotational elastic full sphere. As can be seen from Fig. 2.12, in the general case, on the surface of the full sphere, a series of nodal lines is presented more or less corresponding to parallels and meridians of the Earth's surface. If a seismograph located on the Earth's surface is situated on the nodal line of a certain mode, no records of the oscillations corresponding to this mode are provided.

Lamb (1882) established the slowest oscillation of a spherical surface or the fundamental mode to have one nodal sphere of the R radius with the next oscillation or mode characterised by two nodal spherical surfaces. The root values of their frequency equation are $R \approx 0.8238$ and $r = 0.4792$. The oscillation or mode following them will already have three nodal spheres (R, r, r') and so on.

As the spherical oscillations are decomposed into modes, the sphere is filled with spherical surfaces of various radii. These free oscillation frequencies for an elastic incompressible sphere were also calculated by Gray and Ohringen in 1955 by order of the US Naval Research Department (Purdue Univ. Lafayette Indiana, 1955). These American scientists demonstrated that, for the free oscillations of a sphere, its nodal surfaces are determined by the roots of the corresponding equations of the second, third, etc. order. Thus, the slowest oscillation will have here one nodal sphere of radius R, the next oscillation will already have two nodal spheres of radii R and r and the next oscillation will already have three nodal spheres of radii R, r and r' (Fig. 2.13). The ground-breaking work of Lamb (1882) offers distinct advantages when discussing this issue, since he was the first to determine the most important root values of the equation for the free oscillation frequency of a sphere, which determine the spectral (fractal) structure of its oscillation processes.

Since the displacement for each free oscillation is proportional to a spherical function of a certain order, the designation of each mode will obviously depend on the number of nodal surfaces. Seismologists use the n and l indices to indicate their number in depth and on the surface of the Earth, respectively. The n index refers to the number of nodal surfaces inside the Earth (the centre of the Earth is not included in this number) while the l index is greater by unity, i.e. equal to the number of sectors bounded by nodal lines on the Earth's surface (Fig. 2.14).

Let us first consider the least complex torsional oscillations. Using the n- and l-indices denoting the number of nodal lines, let them be written in the form $_nT_i$. In the same fashion, the n-index refers here to the number of nodal surfaces inside the Earth (the centre of the Earth is not included in this number), while the l-index is greater by unity, i.e. equal to the number of sectors bounded by nodal lines on the Earth's surface.

The simplest torsional oscillation of the $_0T_2$ mode is presented in Fig. 2.14a. Only one nodal surface corresponds to it, cutting the Earth's surface along the equator with the northern and southern hemispheres "twisting" relative to it in opposite directions, with analogous movements observed in the bulk of the Earth. The $_0T_2$ mode represents the fundamental torsional oscillation of the Earth.

The first radial overtone of the fundamental mode T is schematically depicted in the upper right corner of Fig. 2.14b. In this case denoted by $_1T_2$, one nodal line on the surface ($l = 2$) still remains complemented by one n deep (radial) surface with the displacements occurring in opposite directions on its different sides. Figure 2.14b shows that, in this case, the sphere is divided into four regions, with each characterised by a different direction of movement with respect to the neighbouring region.

At this point, let us consider the corresponding designations of the S-modes. The $_0S_2$ mode—as well as the $_0S_0$ mode, often referred to as fundamental when describing

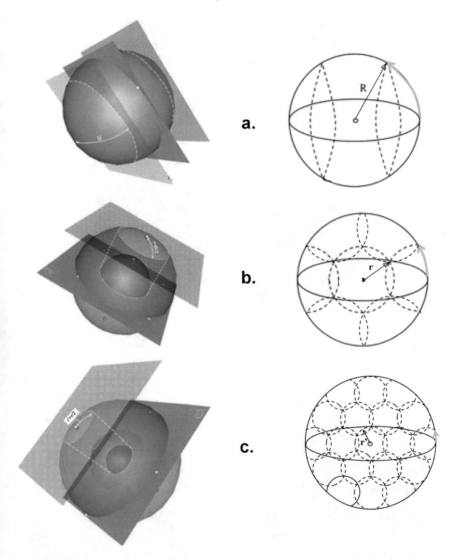

Fig. 2.13 Decomposition of the sphere free oscillations into modes with respect to spherical surfaces of R, r, r' radii Explanations in the text.: the slowest oscillation will have one nodal sphere of radius R, the next oscillation will already have two nodal spheres of radii R and r and the next oscillation will already have three nodal spheres of radii R, r and r'

the free oscillations of a sphere in three-dimensional space, is reminiscent of alternating compression and decompression of an elastic football (Fig. 2.14c). These oscillations correspond to the $n = 2$ s-order spherical function with the sphere oscillating between the prolate and oblate spheroid positions (Figs. 2.14 and 2.15). This type of oscillation also occurs in geophysical problems, for example, when evaluating

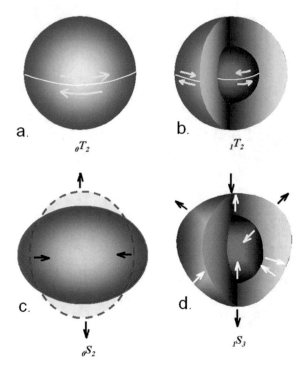

Fig. 2.14 Schematic representation of the $_0T_2$, $_1T_2$, $_0S_2$ and $_1S_3$ modes (according to Bolt 1984)

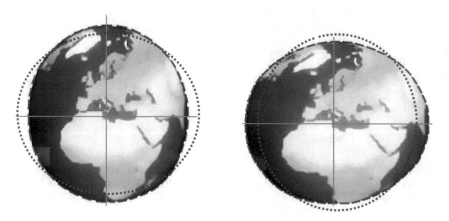

Fig. 2.15 $_0S_2$ Spheroidal mode of the Earth's free oscillations

the amplitudes of oscillations caused by the tidal forces of the Sun and the Moon. They are depicted in the lower left corner of Fig. 2.14c and in Fig. 2.12a in the form of alternating deformations as the fundamental mode of a hollow spherical space. In this case, on the surface, two nodal lines are found coinciding for the $_0S_2$ mode with the parallels of the northern and southern hemispheres (for spheroidal oscillations, the l-index is equal to the number of nodal lines on the surface). Figure 2.12 also demonstrates the two surface overtones of $_0S_3$ and $_0S_4$. It can be clearly seen that, for them, 3 and 4 parallel nodal lines are presented, respectively.

Obviously, as for torsion modes, internal nodal surfaces can also exist here, with the n-index still providing a valid measure of their number. In the lower right corner of Fig. 2.14d, the oscillation scheme of the $_1S_3$ first radial overtone is shown. As can be seen from Figs. 2.12, 2.13 and 2.14, with an increase in n and l, the nodal surfaces approach each other, with numerous seismological measurements indicating that the oscillation frequency of the T and S modes increases simultaneously. Since both n- and l-indices can take values from zero to infinity, the number of possible modes will be much less infinite. However, in the practice of modern seismological observations, only oscillations with values of n and l not exceeding several hundred units are taken into account. In addition, the theory of seismology predicts any deviation from the ideal parameters of elasticity and sphericity to affect various modes of the Earth's resonant oscillations differently. Modes having a very long period (for example, $_0T_2$ and $_0S_2$, which cover the entire planet, are most affected by rotational asymmetry; as the overtone order increases (for example, $_0T_1$ and $_0S_1$), the oscillatory movement is progressively displaced into the upper layers of the Earth, resulting in the intensification of formal ellipticity and tectonic differences.

Figure 2.16 clearly shows how the difference between continents and oceans changes the directions of displacements on the Earth's surface with $_0T_5$ free oscillations. These oscillations are manifested by five displacement zones on the surface, separated by four parallels indicated by dashed lines. On these nodal lines, there

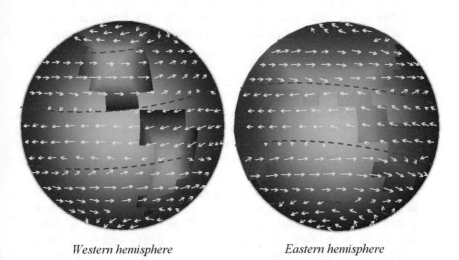

Western hemisphere   *Eastern hemisphere*

Fig. 2.16 Eastern and western hemispheres of the oscillating Earth Continents are shown in grey. Short arrows indicate the direction and amplitude of displacements on the Earth's surface under $_0T_5$ free oscillations (Bolt 1984), as well as the relationship of these oscillations with various types of bulk and surface waves

are no pronounced surface displacements. In the case of the Earth's homogeneous surface, ignoring structural differences between the oceans and continents, the short vectors are parallel to the nodal lines. However, their actual directions differ due to the complex structure of the Earth.

In 1954, when analysing the seismograms of the 1952 Kamchatka earthquake, the leading American experimental seismologist Hugo Benioff identified the phase of a 57-min period with the fundamental spheroidal oscillation of the Earth and, thus, laid the foundation for contemporary research on the Earth's free oscillations.

From 1960 onwards, a large quantity of the Earth's free oscillation spectra from a number of strong earthquakes led to a completely new branch of seismological research, providing an image of our planet not only through a "temporal", but also a "frequency" window. This new area of knowledge dubbed terrestrial spectroscopy turned out to be very closely related to the mechanics of atomic oscillations. In this area, at least, quantum mechanics and seismology converge very closely, with the analogy to atomic scales having demonstrated its worth in formulating and solving the wave equation for the oscillating Earth.

The relationship between the Earth's free oscillations and those of a string can be quite straightforwardly traced by fixing the sphere oscillating relative to the circular nodal line at N and S two polar nodal points and replacing the string with a spherical shell of similar properties (Figs. 2.17 and 2.18). It is in connection with these circumstances that free oscillations of the string and sphere are often regarded as analogues. Using the concepts of projective planes, nodal lines and nodal points, the fundamental mode is quite easy to be imagined in three-dimensional Euclidean space (Fig. 2.18).

If the z axis passes through the centre of the sphere and intersects the spherical nodal surface at two polar nodal points N and S, then the fundamental mode of the oscillating sphere will be represented by the triaxial deformation ellipsoid well known from strength of materials theory (Fig. 2.18).

From the position of three independent observers considering free oscillations of the sphere in the three-dimensional Euclidean space (x, y, z), the displacements of the phase volume relative to the circular nodal line lying in the (x, y) plane will be carried out here in reverse phase with respect to the displacements relative to the circular nodal line lying on the (y, z) plane perpendicular to it. Moreover, the oscillations about the z axis in the (x, y) plane will exactly correspond to the fundamental mode of oscillations for the full sphere surface in the section, which are shown in Figs. 2.12. and 2.14.

It is from the properties of this mapping group that the concepts of the three-dimensional (x, y, z) geometric space for a sphere in Euclidean space are derived. Figure 2.19 clearly demonstrates the idea of three dimensions for a sphere in Euclidean space arising from observations of its projective images, though each of them has only two dimensions; the fact is that they follow each other according to certain laws of projective geometry.

However, how in this case will the oscillations of the sphere be represented on the drawing plane relative to the entire spherical nodal surface of the radius R? And how then can the nodal surface topology of a three-dimensional sphere be imagined in

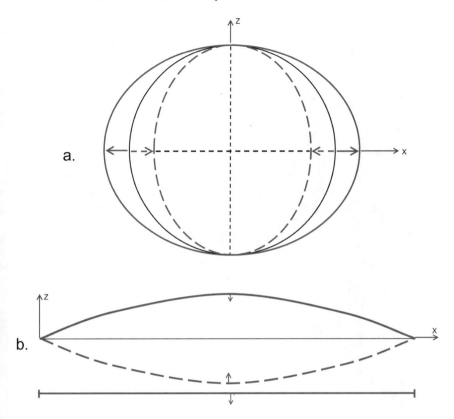

Fig. 2.17 Fundamental mode for free oscillations of a sphere (**a**) and a string (**b**) in projective metrics

terms of a general coordinate system on the layout plane? The difficulty that arises when looking at the sphere is how do we know which space it describes, when at the same time we can see only a small projective part of it?

Henri Poincaré was well aware of the need to expand the dimension of spherical space based on the laws of projective geometry in order to preserve the metric and topological properties of the sphere. According to his ideas, Fig. 2.19 represents the fundamental mode of the sphere free oscillations using the displacement lines of the phase volume with respect to closed concentric nodal lines lying in three mutually perpendicular planes of (x, z), (y, z) and (x, y) on a white screen of two dimensions, i.e. the surface of a paper sheet. Graphically, from the viewing point of three independent observers, the displacements are determined here by mapping the plane into itself (or mapping the plane into a plane), which is often referred to as Poincaré mapping. A closed concentric nodal line here belongs simultaneously to the three planes of (x, z), (y, z) and (x, y), as well as to the drawing plane (Fig. 2.19).

However, even in this case, part of the nodal points and most of the sphere displacements lying in the different projective planes remain completely "invisible" for a

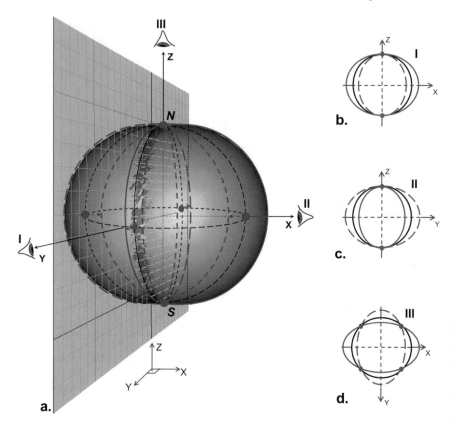

Fig. 2.18 The $_0S_2$ fundamental mode for free oscillations of a sphere in spherical and three-dimensional (x, y, z) Euclidean space (**a**); **b**, **c** and **d** are the fundamental modes of the free oscillations of the sphere with respect to projective nodal lines and S and N projective nodal points lying in three mutually perpendicular planes of (x, z), (y, z) and (x, y) from the viewpoints of three independent observers I, II and III

number of observers. Since these portions lie in the space between the axes and projective planes of the (x, y, z) three-dimensional Euclidean geometry, they require the introduction of a common spherical nodal surface of a radius R or additional spherical coordinates to include them in the process of describing free oscillations of the sphere on the drawing plane (Fig. 2.18).

In order to solve this problem and expand the dimensionality of the spherical space under study when mapping it on the drawing plane while preserving its topological and metrical properties, we will require a powerful mathematical apparatus of elliptic or Riemann geometry in order to precisely study the properties of multidimensional "surface" spaces (n-dimensional surfaces in a space of $n + 1$ dimensionality).

Obviously, as in projective Riemannian geometry, the projective plane can in this case serve as the topological model for the spherical phase space. This plane is obtained from the usual Euclidean plane by joining an ideal (infinitely distant)

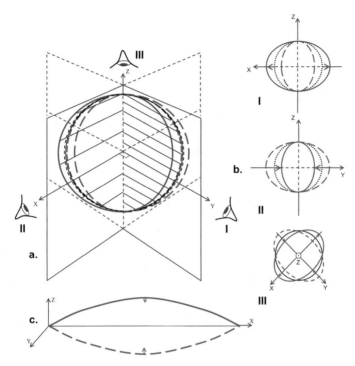

Fig. 2.19 The fundamental mode of free oscillations of a sphere in the form of a deformation ellipsoid in (x, y, z) three-dimensional space. The mode is represented by phase volume displacement lines of relatively concentric nodal lines and nodal points lying in three mutually perpendicular planes of (x, z), (y, z) and (x, y) by the mutual mapping of these planes into each other (**a, b**). Displacements are represented as an integrated scheme simultaneously seen by several observers $(x, y$ and $z)$. The fundamental mode of a vibrating string (**c**)

point and an ideal straight line: each line is supplemented by one infinitely distant point ($+\infty$ and $-\infty$ are identified and referred to as a point) being the same point and different ones for parallel and non-parallel lines, respectively. All these connected points lie on one infinitely distant perspective line also joined to the plane. On the one hand, the metric properties of the sphere will in this case correspond to the infinitely distant perspective point, while, on the other hand, they will coincide with the metrical properties of an ordinary sphere. More precisely, for any point of the projective plane, there is a neighbourhood isometric to some part of the sphere having a radius identical for the entire plane of a given Riemannian space and equal to its radius of curvature.

Thus, instead of the notion for an ordinary surface in projective Riemannian geometry, the notion of a manifold appears. Moreover, on a G projective plane, each pair (a neighbourhood of a point and a homeomorphism, or its mapping into R^n, preserving its topological properties, i.e. compactness, connectedness, etc.) is referred to as a manifold map with the collection of maps covering all the manifold presenting the atlas of manifold.

Let us recall that it was according to the concept of "homology" that Poincaré, thanks to his ingenious intuition, saw the pivot for the entire further development of topology. Moreover, the first formulation of this concept (in his 1895 memoir *Analysis situs*) appealed precisely to its direct geometric visibility, under which, only a few years later, a strictly logical basis would be developed. This feature of topological solutions, associated with the features of perspective and projective perception of the world around us through human eyes, has also been widely used in our work.

Let us try to explain the abstract mathematical generalisations of new properties and dimensionality of multidimensional spaces using particular graphic examples. From the standpoint of Riemannian geometry, a graphic model (manifold map) of the fundamental mode for an oscillating sphere on the projective plane G can be obtained as follows. Let us take a sphere and cut it with a G plane passing through its centre (Fig. 2.20a). Then, using the concepts of manifold introduced above, three mutually perpendicular planes of (x, z), (y, z) and (x, y) are attached to it, as shown in Fig. 2.20a. Obviously, traces from a section of the R-radius sphere by three mutually perpendicular planes passing through the nodal points on its surface will correspond to arcs of a 2R-radius circumference. At the same time, in the Riemannian (spherical)

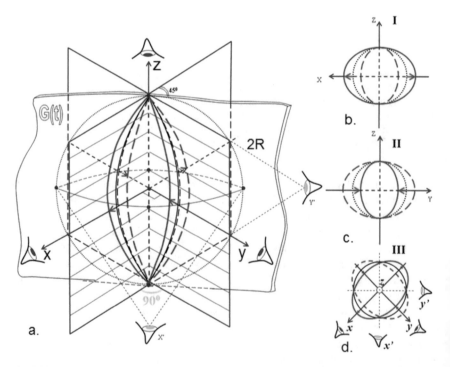

Fig. 2.20 Topological map (manifold map) of the fundamental mode for an oscillating sphere on the $G(t)$ projective plane (**a**); **b, c** and **d** are the fundamental mode (deformation ellipsoid) in the (x, y, z) three-dimensional space; the oscillations are shown relative to three mutually perpendicular planes

geometry, the role of the straight lines is played by the "big circles" or circumferences obtained by the section of a sphere by any plane passing through its centre. Thus, the dashed straight lines shown in Fig. 2.20 passing through the centre of the sphere, on the one hand, can be obtained by cutting the sphere with three mutually perpendicular planes of (z, x), (z, y) and (x, y) and, on the other hand, the circumference lying in the G projective plane belongs to these lines. The circumference projections belonging to (z, x) and (z, y) two mutually perpendicular planes are shaded in Fig. 2.20a.

Being attached to the projective plane G, these projections form arcuate curved lines on it, while the projections of the lines lying on different hemispheres coincide and are therefore denoted by double (solid and dashed) lines in Fig. 2.20a. If we return again to Fig. 2.19, the deformation ellipsoid shown there in three-dimensional Euclidean space (z, x, y) using (z, x), (z, y) and (x, y) three mutually perpendicular planes can be represented on the G projective plane of Riemannian geometry in the form of spatially aligned arc-shaped displacements also belonging to the same three mutually perpendicular planes of (z, x), (z, y) and (x, y), or, rather, to their projections on the projective plane G (Fig. 2.20).

In fact, already in Fig. 2.19a, by trying to depict the ellipsoid's oscillations in three-dimensional space and preserve the topological and metric properties of the spherical space necessary for graphical construction of the circular nodal line forming the general coordinate system, we were forced to resort for mapping two mutually perpendicular planes of (z, x) and (z, y) one to each other. Figure 2.20 shows how these mappings can be developed within the framework of Riemannian geometry, as well as to map not only two mutually perpendicular planes to each other, but also to obtain their mapping as refracted through a spherical nodal surface onto the projective plane G. Moreover, Riemannian geometry provides for strict preservation of the topological and metric properties of three- and two-dimensional spherical space necessary for such topological (graphical) solutions.

Thus, as Henri Poincaré had done before by reflecting the plane into itself, we obtained an image of a spherical figure of three dimensions on a spherical screen of two dimensions and thereby combined the three-dimensional Euclidean geometry with spherical Riemannian geometry. Although, in his famous work *Science and Hypothesis*, Poincaré noted that "…this task is extremely simple for the geometrician…", it didn't seem so to us. Indeed, for this purpose, we had to turn to Riemannian geometry and obtain the ability for mapping (or "folding") a three-dimensional Euclidean representation of the fundamental mode of an oscillating sphere in a two-dimensional spherical Riemannian space using three mutually perpendicular planes attached to the projective plane G.

At the same time, along with three blue-eyed observers of (x, y, z), who can see the fundamental mode for free oscillations of the sphere in three-dimensional Euclidean geometry, three green-eyed observers of (x', y', z') appear able to see the same triaxial deformation ellipsoid (the fundamental mode of the sphere) in the coordinates of spherical Riemannian geometry (Fig. 2.20). From the standpoint of contemporary mathematics, these mappings—or superimposing—of the topological and metric properties for spherical and Euclidean spaces are not only a further Riemannian development of the idea for plane mapping into itself (or Poincaré map) already considered above, but also comprise a mapping of the surface onto the plane within a perspective and projective measurement scheme, which is associated with the name of the American mathematician of Whitney (1955) and is referred to as the Whitney singularity theory of smooth mappings. Moreover, in the framework of Riemannian geometry, these mappings can be considered as a far-advanced theory of nonlinear dynamical systems—a grandiose "maxima and minima" generalisation of the function study, where the functions are replaced by their topological maps of spaces with varying positive and negative curvature, taking into account the perspective and projective perception of the world through the human eyes. However, as Poincaré noted in his fundamental work *Analysis situs*, the question immediately arises: "… is it necessary to replace the language of analytical research with the language of geometry, which loses all its advantages as soon as the opportunity to use the senses disappears?" However, this new language turned out to be more accurate. In addition, by analogy of varying negative and positive curvature spaces it is possible to use ordinary geometry to create associations of fruitful ideas and suggest useful generalisations (Poincaré 1974).

In this regard, one cannot fail to notice that the mapping of a spherical surface onto a plane in Whitney's theory evolves a mapping of a plane point to each point on the surface (Fig. 2.21). If the surface points are given by the (x_1, x_2) coordinates on the surface and the plane point—by the (y_1, y_2) coordinates on the plane, then the mapping is obviously set by the pair of $y_1 = f_1(x_1, x_2)$, $y_2 = f_2(x_1, x_2)$. functions. Here, a map is called smooth if these functions are smooth (i.e. differentiable enough times, for example, polynomials).

From the view of projective metrics, mappings of smooth surfaces onto a plane surround us from all sides. Indeed, as we were able to see, the visible contours of the sphere are the projections of its bounding surfaces on the retina. Looking more closely at the bodies around us, Hassler Whitney studied the features of these visible contours. He noted that, in cases of general situation, there are features of two types. As a result of free oscillations, all other features are destroyed even with a small perturbation of bodies or mapping directions, while the features of these two types are stable and preserved under small mapping deformations. An example of a feature of the first kind, the Whitney fold, consists of a feature occurring when a sphere is projected onto the G projective plane at the equator points (Fig. 2.21b). At suitable coordinates, this mapping is defined by the formulas of the $y_1 = x^2_1$, $y_2 = x_2$ quadratic equation and has the shape of a circle on the projective plane of the drawing. When describing free oscillations of a spherical space, this feature implies the transition from one region of structural stability to another, to be carried out here through a non-rough state corresponding to the bifurcation value of parameter b in Fig. 2.22.

Since we have already encountered these bifurcation phenomena when describing free oscillations of a string (Fig. 2.4), we know that the behaviour of the system changes stepwise as a result of such a transition. The psychological reflection of such a transition with an arbitrarily small change in the parameter consists in the term "catastrophe" included as a keyword in the name of the "catastrophe theory" mathematics section studying the structural features of free oscillations for the limiting points of space in the form of the set of algebraic and/or transcendental equation roots

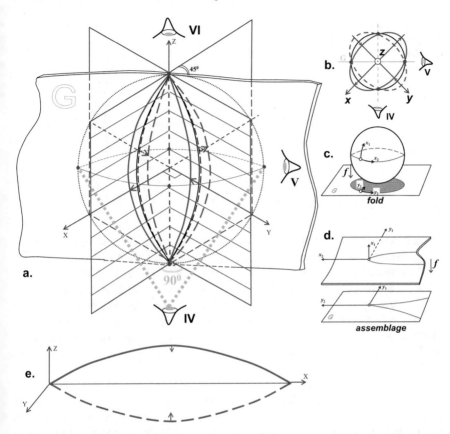

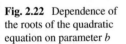

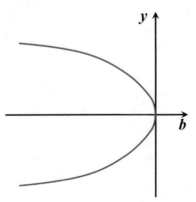

Fig. 2.21 Fundamental mode of an oscillating sphere on the projective plane. The arrows show the longitudinal displacements of compression-tension along the axial lines of the ellipsoid, the solid and dashed lines indicate the volume deformation in the Riemannian (**a**) and Euclidean (**b**) geometry. Fold (**c**) and cusp (**d**) of the sphere's surface on a plane. Further explanations in the text

Fig. 2.22 Dependence of the roots of the quadratic equation on parameter b

Fig. 2.23 Dependence of
the $y_3 + b_1 y + b_0 = 0$ cubic
equation roots on b_1 and b_0
parameters

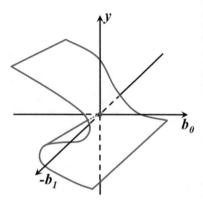

that arises every time when trying to describe the propagation of wave disturbances using systems of rays or wave fronts in projective space (Arnold 1990).

The considered example exactly reflects the simplest catastrophe called the "fold", since the $b = 0$ bifurcation point resembles a section of the fold image on a certain surface by the plane. In general cases, mapping the sphere surfaces on the retina has just such a feature: here, there is nothing surprising, since we can look at a glass ball or globe from either side. What is surprising is that, in addition to this feature (fold), we find exactly one more feature everywhere, but almost never notice it (Arnold 1990).

This second feature is called the "Whitney cusp". In the general case, it is obtained when projecting onto the surface plane depicted in Figs. 2.23 and 2.24. This surface is given by the formula $y_1 = x^3{}_1 + x_1 x_2$ in the space with (x_1, x_2, x_3) coordinates and projected onto the (x_2, y_1) projective plane, while the mapping is specified in local coordinates by the formulas of the cubic equation of $y_1 = x^3{}_1 + x_1 x_2$, $y_2 = x_2$ (Arnold 1990).

On the projective plane of the drawing, the "cusp" stands out as a semi-cubic parabola with a cusp point at the origin. This curve divides the projective plane into two parts: smaller and larger. The points of the smaller part have three inverse images (three points of the sphere surface are projected into them). The points of most of the circular nodal line (folds) and the points of the curve (double nodal line or a cusp) have only one and two inverse images at a time, respectively. When approaching a curve from a smaller part, two inverse images (of three) merge and disappear (in this position, the feature is presented by a fold). In the case of approaching a cusp (nodal point), all three inverse images merge.

Thus, at the intersection of the curves, the multiplicity of root degeneracy increases by one to reach the maximum possible value for this equation. This point is called a cusp catastrophe. The meaning of the last term becomes clear in the case of depicting the dependence of the equation roots on the parameters. Since, unlike the case of the quadratic equation, the number of parameters is equal to two, this dependence is depicted by some surface in three-dimensional space (y, b_0, b_1) in Fig. 2.23.

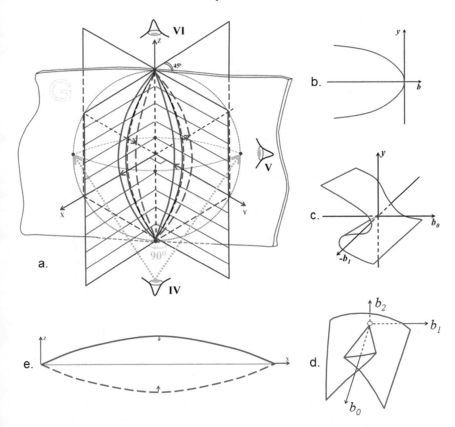

Fig. 2.24 Fundamental mode of an oscillating sphere on the *G* projective plane. The arrows show the longitudinal displacements of compression-tension along the axial lines of the ellipsoid, the solid and dashed lines indicate the (**a**) volume deformation in the Riemannian geometry. Fold (**b**), cusp (**c**) and "swallow tail" (**d**) in projecting the sphere surface onto the G projective plane. The fundamental mode of a vibrating string (**e**). Further explanations in the text

The fold lines are clearly seen to converge at the cusp point (Figs. 2.24 and 2.25). It is the characteristic form of deformation of the $y = y(b_0, b_1)$ surface in the neighbourhood of this point that defined the name of the catastrophe in the case under consideration, which determines the relationship of three-dimensional Euclidean geometry with b_0 and b_1 two bifurcation parametric axes of projective spherical geometry in describing free oscillations of the sphere (Fig. 2.24). Whitney showed the cusp to be stable in the sense that every near mapping has a similar feature at a suitable near point (i.e. such a feature that, as a result of oscillations in suitable coordinates in the neighbourhood of the specified point, the deformed mapping is written in the same formulas as the original map in the neighbourhood of the starting point).

In accordance with the dynamics of the vibrating string, we also can obtain graphical solutions of these equations for the sphere. Indeed, if, within the framework of

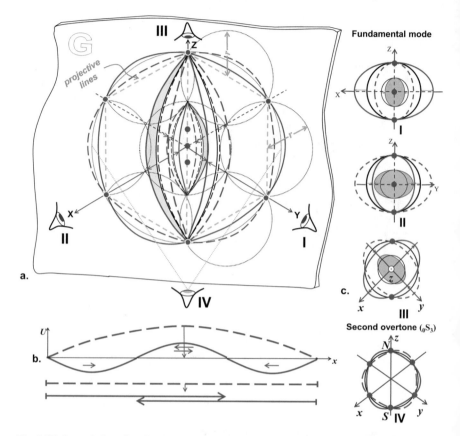

Fig. 2.25 Interrelation of oscillation and wave processes in three-dimensional Euclidean and spherical Riemannian spaces in comparison with a vibrating string. The second overtone is shown for the string and sphere. If the nodal points of the second overtone for an oscillating sphere are connected by straight lines on the G projective plane, the nodal lines are obtained with the corresponding circular nodal lines of radius r in the Riemannian space

the Euclidean and Riemannian geometry axioms, the nodal points of the second overtone for an oscillating sphere are connected by straight lines on the G projective plane, the nodal lines are obtained with the corresponding circular nodal lines of radius r in the Riemannian space (Fig. 2.25).

Upon subsequent decomposition into modes with respect to these spherical nodal surfaces of radius r, nodal points are formed again with new nodal surfaces of radius r' capable of being obtained in the case of connecting these points (Fig. 2.26).

In this case, when a two-dimensional 2D spherical space oscillates relative to a three-dimensional 3D Euclidean one with (x, y, z) coordinate axes of the latter intersecting at a nodal point in the centre of the sphere, the sphere is bundled into surface shells oscillating in reverse phases, i.e. an outer and inner core (Figs. 2.25 and 2.26). On the other hand, in accordance with the dynamics of the oscillatory process,

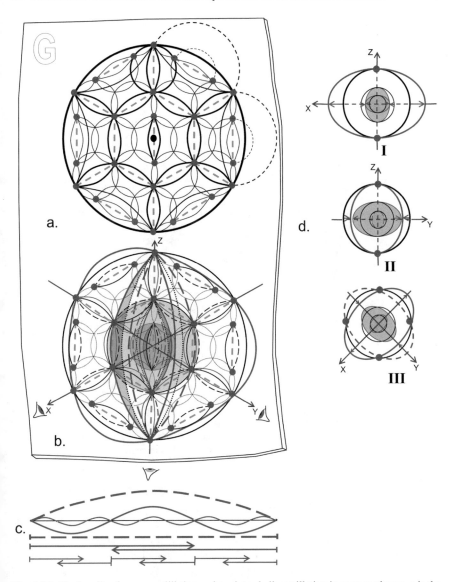

Fig. 2.26 The bundle of a non-equilibrium sphere into shells oscillating in reverse phases. **a** is the occupation of a two-dimensional spherical volume with a self-similar spatial grid; **b** is the formation of displacements relative to the spherical nodal surfaces of this grid; **c** is the correspondence of displacements in a non-equilibrium sphere to the same in a non-equilibrium string; **d** is the reverse phase oscillations of the embedded shells from the position of three independent observers

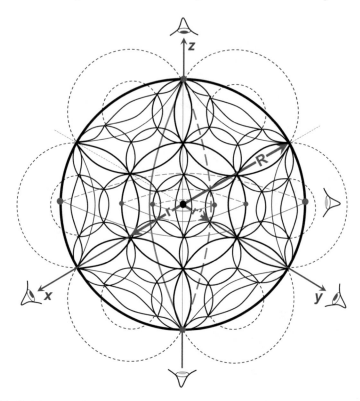

Fig. 2.27 Nodal spherical surfaces and singular points for the free frequencies (or modes) of an oscillating sphere on the Riemannian geometry projective plane of R, r and r' radii. The slowest oscillation of a spherical surface has one nodal sphere of R, the next oscillation or mode has two nodal spheres of R and r, while the next oscillation or mode has three nodal surfaces of R, r and r'

the sphere is decomposed into 2D spherical nodal surfaces of (R, r, r') various radius with the oscillations of three-dimensional phase volume occurring about them.

In three-dimensional Euclidean space, newly-appearing deformation ellipsoids are an exact, half-reduced copy of the fundamental mode representing a regular fractal. Evidently, the formation of this fractal is associated with (x, y, z) three-dimensional deformation processes and its dimensionality on the projective plane G is determined by filling the phase space with self-similar volumes, i.e. spheres with the radius of each subsequent one being half the radius of the previous (Figs. 2.25, 2.26 and 2.27).

For each isolated spherical phase volume newly formed in this way, its own (z, x, y), $(z_{1-6}, x_{1-6}, y_{1-6})$ and $(z_{1-18}, x_{1-18}, y_{1-18})$ three-dimensional coordinate system of Euclidean geometry can be distinguished; the displacements of the phase volume can be depicted in the form of isolated deformation ellipsoids relative to spherical nodal surfaces (Fig. 2.28).

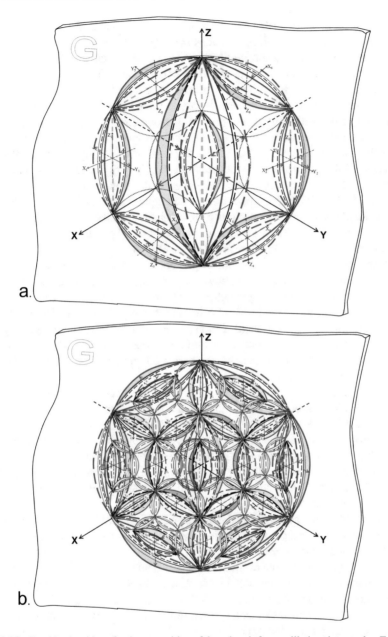

Fig. 2.28 Graphic algorithms for decomposition of the sphere's free oscillations into modes. Explanations in the text.: for each isolated spherical phase volume newly formed, its own (z, x, y), $(z_{1-6}, x_{1-6}, y_{1-6})$ and $(z_{1-18}, x_{1-18}, y_{1-18})$ three-dimensional coordinate system of Euclidean geometry can be distinguished; the displacements of the phase volume can be depicted in the form of isolated deformation ellipsoids relative to spherical nodal surfaces

This method for describing free oscillations of a sphere in three-dimensional Euclidean space not only clearly demonstrates the isolation of the outer shell, with the outer and inner nucleus oscillating in reverse phases relative to the central nodal point, but also characterises the propagation of oscillations in time and space. Moreover, as we approach the projective plane, the sizes of spherical nodal surfaces successively filling it will decrease almost to minus infinity—∞, with the set of elementary deformation ellipsoids, presenting exact three-dimensional copies of the fundamental mode in three-dimensional Euclidean space, becoming almost infinite. With this approach to the image of the fractal dimensionality of spherical nodal surfaces, all their projections (manifolds) on the $G(t)$ projective plane—and hence the manifolds of elementary ellipsoidal (spheroidal) displacements—can be studied as submanifolds of Euclidean space using a technique similar to differential geometry.

Realising this filling within limits of infinity, the intersection of spherical nodal surfaces can be seen to form a system of nodes (or nodal points) and organise dense hexagonal packing in multidimensional space of $G(t)$.

Within the framework of Euclidean geometry spaces, such decompositions of oscillations represent an exact analogy to expansion in Fourier series, when, within the framework of Euclidean geometry, complex wave oscillations can be considered as a superposition result of simple harmonic components, each having its own frequency, wavelength and phase. Hence, as in the case of a vibrating string, the oscillating sphere is equivalent to an infinite number of independent pendulums, i.e. a discrete set of free frequencies arises in a continuous problem describing the oscillations of a sphere.

In fact, in a series of topological maps (Figs. 2.29 and 2.30), when creating a bifurcation diagram of free oscillations of the sphere relative to the axes of three-dimensional Euclidean geometry and a grid of nodal spherical surfaces (or their maps on the projective plane), the graphical method is practically implemented proposed by mathematicians to calculate the fractal dimensionality (or capacity) of sets. In the framework of Euclidean geometry, this method is based on sequential filling of the projective plane (G) with additional spherical nodal surfaces of various diameters by forming a dense packing.

When taking this approach to the description of an oscillating sphere, as in the case of a vibrating string, a graphical solution of the wave equation is easily obtained in the form of a standing wave. However, in this case, the most important part of the bifurcation dynamics for the oscillatory process eludes us even through the infinite-dimensional mesh of the fractal network, despite the latter providing us, within the framework of Euclidean geometry, with the spatial decomposition of the spherical oscillatory process into its elementary three-dimensional components, unprecedented for contemporary natural science.

On the other hand, if we, using the experience of describing free oscillations of a string, combine the nodal points of the vibrating string with the N and S nodal points at the poles of the sphere and its centre, as shown in Figs. 2.29 and 2.31, then the relationship of the spheroidal and radial oscillations of the first $_0S_2$ and second $_0S_3$ sphere overtones can be found in comparison with the second and first modes of a vibrating string.

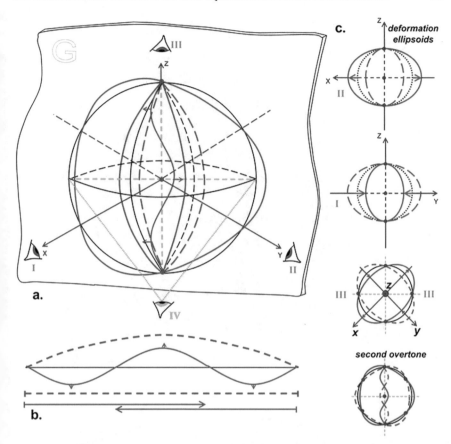

Fig. 2.29 Fundamental mode and second overtone of a sphere oscillating in three-dimensional Euclidean (observers I, II and III) and spherical Riemannian (observer IV) spaces in comparison with a vibrating string. The second overtone is shown for a string and a sphere. In spherical space, the second overtone is realised with respect to the nodal line passing through the centre of the sphere in the form of radial oscillations and their inverse images arising with respect to the nodal circular line, or spheroidal oscillations

Looking at Fig. 2.31, it is easy to infer that the first overtone $_0S_2$ in a spherical projective space implies an exact analogy (or inverse image) of the fundamental mode in (x, y, z) three-dimensional Euclidean space. As a standing wave, $_0S_2$ mode forms nodal points at the poles and equator in the intersections of a closed nodal line and mutually perpendicular axes of (z, x). Since this mode ($_0S_2$) has a longer wavelength than the second overtone mode ($_0S_3$) preceded it, it was taken as the fundamental mode in classical Newtonian mechanics.

The dynamics of free oscillations of the sphere was reduced to a sequential wavelength expansion into Fourier series and, thereby, as in the case of a vibrating string, the basic idea of the wave process—i.e. the dual nature of the wave properties of

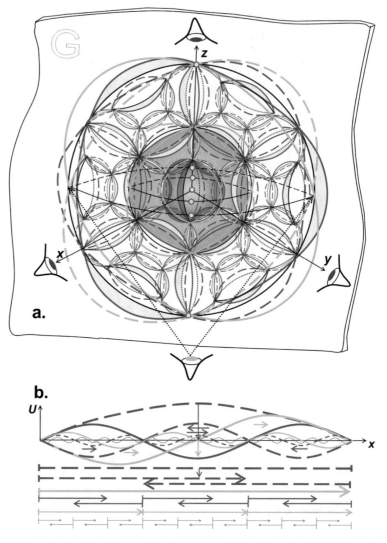

Fig. 2.30 Interrelation of oscillatory and wave processes in three-dimensional Euclidean and spherical Riemannian spaces (**a**) in comparison with a vibrating string (**b**). The first, second, third, fourth and fifth overtones (modes) are shown for a string and a sphere

matter and the bifurcation dynamics of the relationship between standing and travelling waves—was excluded from it. Thus, from the standpoint of projective measurement in an oscillating sphere, as well as in a vibrating string, the splitting of oscillation and wave processes is established. Contrary to the sequence of reading spectrograms in physics and mathematical expansion in Fourier series, the first overtone does not

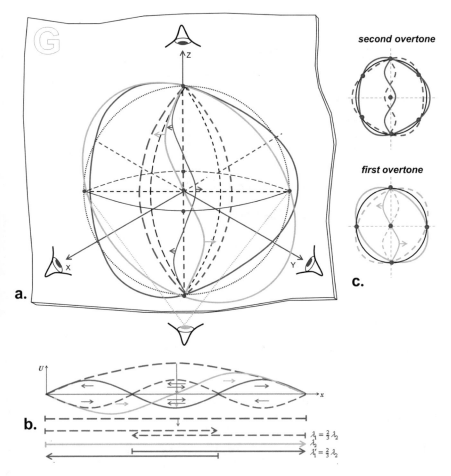

Fig. 2.31 Bifurcation algorithms for the ratio of the second and first overtones in a sphere oscillating in three-dimensional Euclidean (observers I, II and III) and spherical Riemannian (observer IV) spaces in comparison with a vibrating string. The fundamental mode, the first and the second overtones are shown for a string and a sphere. In spherical Riemannian space, the second and the first overtones are realised with respect to the nodal line passing through the centre of the sphere in the form of radial oscillations and their inverse images, spheroidal oscillations, arising with respect to the nodal circular line

precede the second, but occurs later as a result of bifurcation. These bifurcation relations of various modes in an oscillating sphere determine the inextricable relationship of standing and travelling waves in nature.

The following figures demonstrate the ratio of different modes for the Earth's free oscillations (Fig. 2.30, 2.32 and 2.33).

In this case, as can be seen from Figs. 2.30 and 2.34, in three-dimensional Euclidean space relative to the (x, y, z) coordinate axes intersecting in the centre

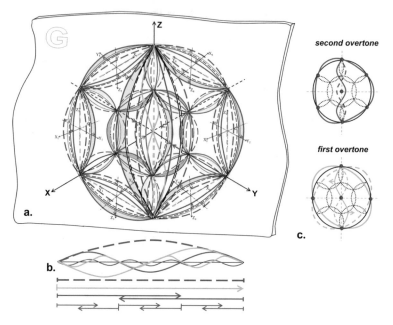

Fig. 2.32 Ratio of the fundamental mode for the second and the first overtones of a vibrating string and a sphere (**a**, **b** and **c**). The ratio of the radial and spheroidal modes for free oscillations of a sphere (**d**)

of the sphere, it is bundled into a surface shell oscillating in reverse phases, i.e. an outer and inner core.

The bifurcation mechanism of the ratios for the second and the first, the fourth and the third overtones of spheroidal, radial oscillations is shown on the right-hand side of Figs. 2.34b and 2.30. Here, in the framework of the combined coordinate systems of three-dimensional Euclidean and projective spherical geometry, we may describe these modes not only relative to the central point of the coordinate system located in the centre of the sphere, but also relative to special nodal and bifurcation points on any of its surfaces and in its entire volume.

Considering this problem from the perspective of topology, we will take some spherical surface and consider it as a rubber film that is both compressible and stretchable, but not tearable, which corresponds to conservative Newtonian systems from the perspective of topology. According to these restrictions, none of the previously allowed geometric operations in three-dimensional Euclidean space can transform a sphere into a torus (donut). However, it can be very easily done in the projective spatiotemporal coordinates of relativistic dynamics by following the decomposition of the oscillatory and wave process into modes relative to the "folds" and "cusps" to project a sphere onto a plane (Fig. 2.35).

The superposition result for the main oscillations of a sphere relative to the "fold" forms toroidal streams on the surface of the sphere, which are topologically linearly

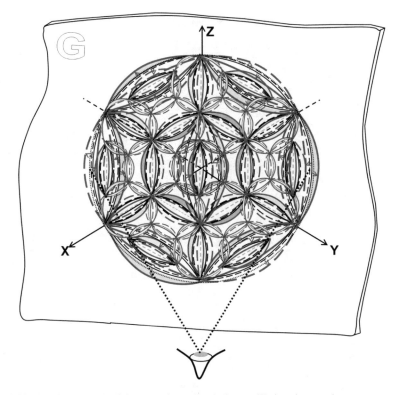

Fig. 2.33 Consistent result of decomposing sphere's free oscillations into modes

elongated roll-like structures and physically realise the tipping phenomenon in a density-unstable spherical shell (Figs. 2.35. and 2.36).

These structural forms on spherical surfaces, associated with dissipation (or energy dissipation), were obtained both by us and other researchers in numerical and physical experiments (Fig. 2.36).

Such spatiotemporal mappings of points, lines and planes into themselves are often called Poincaré mappings and turned out to be connected with the solutions for systems of three ordinary partial differential equations connecting the functions $x(t)$, $y(t)$, $z(t)$, like the solutions of the d'Alembert wave equation for a vibrating string (Fig. 2.37).

Thus, in the resonant combination of discrete point systems with distributed wave function, qualitatively new discrete-wave types of ordering arise along with associated structural and morphogenesis processes. The movement of each individual particle occurs here along individual closed trajectories (or streamlines) with respect to special nodal points, lines and nodal surfaces.

Therefore, if we associate a family of streamlines with a family of displacement lines characterising a set of modes for free oscillations of a sphere and various types of bulk and surface waves, then, in this case, the entire phase volume is divided

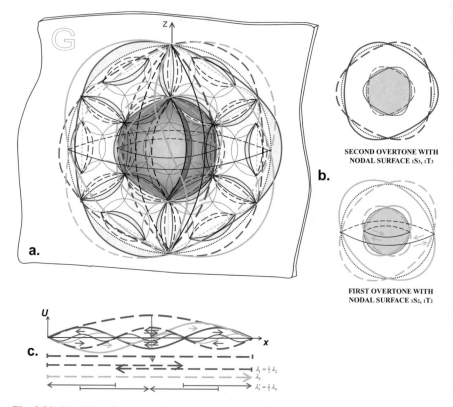

SECOND OVERTONE WITH NODAL SURFACE $_1S_3$, $_1T_3$

b.

FIRST OVERTONE WITH NODAL SURFACE $_1S_2$, $_1T_2$

Fig. 2.34 Bundling of a non-equilibrium sphere into embedded shells and the analogy with a vibrating string

into spherical surfaces with boundaries presenting the equilibrium levels with thermodynamic oscillations of a particle ensemble occurring with respect to them in non-equilibrium in terms of density medium.

Along with the roll-like structures, depending on the asymmetry sign (ε), hexagonal cells are optimally manifested either by emersion in the centre and submerging in the L-hexagon edge parts, or submerging in the centre and emersion in the G-hexagon edge parts. It is worth noting that there is always an area where these two types of hexagonal cells are stable with two-dimensional rolls.

In addition to the hierarchy of closed trajectories corresponding to periodic oscillatory motions, trajectory hierarchies with vortex and spiral components are outlined here. Hierarchies of ring structures arise. The process of such discrete-wave fragmentation of the displacement lines and branching of phase trajectories in a two-dimensional phase space can continue indefinitely. In this case, the motion of matter along the phase trajectories after one or two bifurcations acquires an irregular chaotic structure. In Figs. 2.39 and 2.40 any point is simultaneously on two or three phase

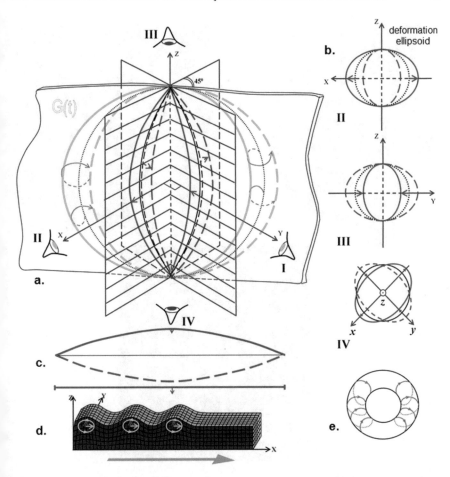

Fig. 2.35 Result of the Poincaré resonance interactions between the sphere's free oscillations represented on the G projective plane by the displacement lines of the fundamental mode and the elliptical trajectories of individual particles, represented by streamlines (**a**). The displacements of the fundamental mode in the Euclidean coordinate system (**b** and **c**). Comparison of a vibrating string with the fundamental mode (**d**). Roll-like structures on the surface of the sphere (**e**)

trajectories with the direction of its motion impossible to be predicted even in a two-dimensional phase space at any given time.

Such behaviour of streamlines indicates the occurring diffusion of discontinuities or fractions in the motion scheme of a single point on the phase diagram. With respect to the motion of each individual particle, the situation is analogous to a random or Brownian motion. At each step, the future of the particle is undetermined. The similar nature of the relation between the discrete-wave nature of the propagating energy (or the distribution of impulses) and the behaviour of a single particle in unstable in terms of density systems poses the same problem that the creators of quantum mechanics encountered in studying the microworld at the turn of the nineteenth and

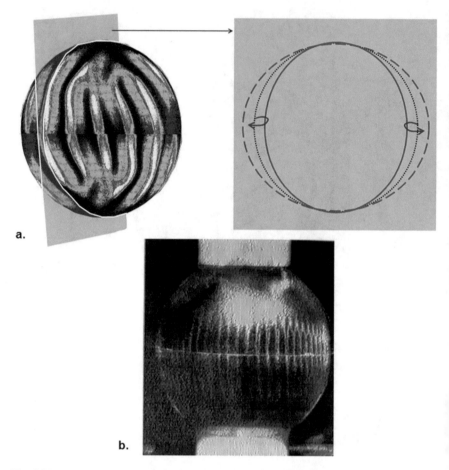

a.

b.

Fig. 2.36 Mathematical (**a**) and physical (**b**) modelling of roll-like dissipative structures on the surface of a density-unstable sphere as a result of implementing the oscillatory motions with an increase and decrease in the phase volume relative to the nodal line. **b**—Convection in a rotating spherical shell with an inner radius of $R_0 = 5$ cm and a distance between the inner and outer surfaces of $d = 0.32$ cm. The structure of convective motion is observable due to the use of a suspension containing lamellar particles able to line up in linear formations under the conditions of shear displacements

twentieth centuries: the impossibility of simultaneous determination and description of the coordinate and impulse for each individual particle. The introduction of wave functions into the phase space introduces elements completely alien to the local description at the level of trajectories, based on the laws of classical Newtonian mechanics. The description in terms of trajectories becomes insufficient and, in some cases, also impossible, since initially arbitrary close trajectories exponentially diverge in time (Figs. 2.39 and 2.40).

Fig. 2.37 Scheme for
constructing a Poincaré
mapping

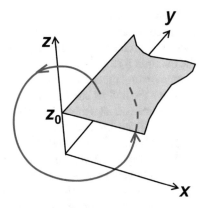

Thus, having considered various methods of physical and mathematical description of the natural processes for spontaneous structuring of unstable in terms of density masses, regardless of the dynamic or thermodynamic causes entailing this instability, we have revealed their discrete-wave nature. Reflecting the main property of the wave oscillatory process, i.e. its mode decomposition, it turned out that the wave properties of matter in the processes of spontaneous structuring of unstable in terms of density masses reveal themselves in the formation of the fractal (or fractional) dimensionality of the space under study. Formation processes repeat themselves in different scales, resulting in appearance of self-similar structures at different scale levels. When taking such an approach towards understanding density (or convective) instability phenomena, they can be characterised using the concepts of amplitude, phase, wavelength, period and frequency. The change in the state of such systems acquires a collective character with questions concerning individual characteristics of distinct elements set to one side. In the description of such systems, the wave properties of matter acquire a decisive importance.

Unlike the equations of the trajectories in classical Newtonian mechanics, traditionally used in describing the phenomena of spontaneous structuring of masses in the non-equilibrium distribution of densities, the introduction of wave functions into the phase space leads to wave equations. These equations are partial differential ones, since, along with the time one, they contain coordinate derivatives and mean a transition from a set of oscillation motions for individual material points to continuous media. As we have seen, this duality in the description of the wave properties of matter manifests itself whenever attempts to expand the dimensionality of the phase space are made and serves as a direct reflection of the wave dual nature. According to the figurative expression of the Australian-born seismologist Bolt (1984), travelling waves are the inhabitants of the world observed through a time window; standing waves are the creation of a frequency window, going hand in hand with spectral representations. Obviously, in the full description for the manifestation of the discrete-wave properties of matter, both these complementary worlds of time and frequency, revealed in all-natural processes from the standpoint of the bifurcation relationship of standing and travelling waves, are necessary to be used.

From the position of topology, the sequential decomposition into modes of the R, r, r' radii occur here with respect to three pairs of nodal points on the surface of the sphere and relative to the nodal point in its centre (Fig. 2.41). In the same sequence, the maximum amplitudes of the radial, spheroidal and torsional displacements of the corresponding modes are realised.

Since the hollow shell of the Earth's surface would be referred by topologists as a two-dimensional sphere, an ordinary school globe can be used for greater clarity when considering the Earth's structuring and morphogenesis processes (Figs. 2.32, 2.33, 2.34, 2.35, 2.36, 2.37, 2.38, 2.39, 2.40, 2.41, 2.42, 2.43 and 2.44).

Such an approach is particularly appropriate when studying the Earth's structuring and morphogenesis processes; for example, it was just such a careful consideration of the peculiarities for the location of oceanic depressions and continents that led Alfred Wegener to the important theory of continental drift.

Obviously, in order to describe the sequence of sphere's free oscillations, we must either move relative to three pairs of nodal points, or use a system consisting of three independent projective observers (Figs. 2.41, 2.42, 2.43 and 2.44).

In fact, as we already know, the distribution for cellular structures of thermogravitational instability on the Earth's surface is the formulation of the famous d'Alembert wave equation in the new projective coordinate system presented in the form of "folds" and "cusps" on the projective Riemannian plane (Figs. 2.42, 2.43 and 2.44). Observer IV, who examines a sphere of radius R relative to the nodal line of the "fold" and special nodal and bifurcation points in the centre of the sphere and on its surface, will see radial, spheroidal and torsional oscillations corresponding to this spherical surface.

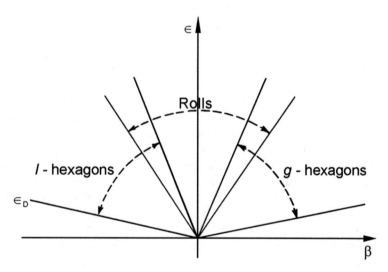

Fig. 2.38 Stable regions of convective instability in the form of rolls and L- and G-hexagonal cells as a function of ε amplitude and β asymmetry parameter (Busse 1978)

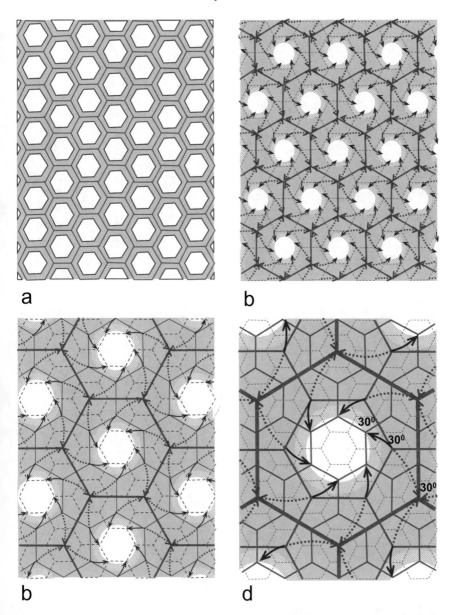

Fig. 2.39 Dynamics of the *l*-hexagon cellular structures (emersion in the centre, submersion in the marginal parts). The figure shows the dynamics of the "respiratory rhythm" or the inversion of *l*- and *g*-hexagons of different generations with the formation of vortex and ring structures

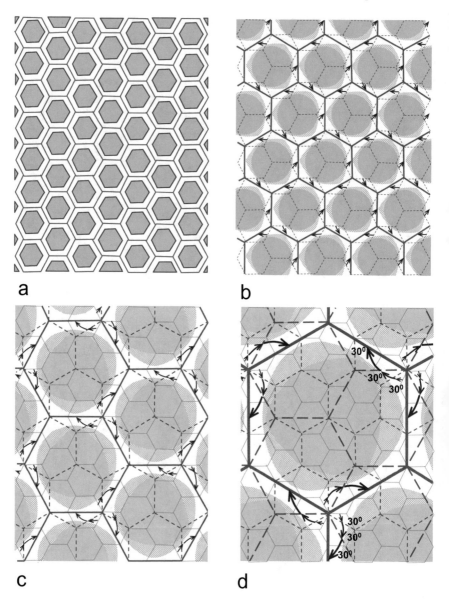

a

b

c

d

Fig. 2.40 Dynamics of the g-hexagon cellular structures (submersion in the centre, emersion in the marginal parts)

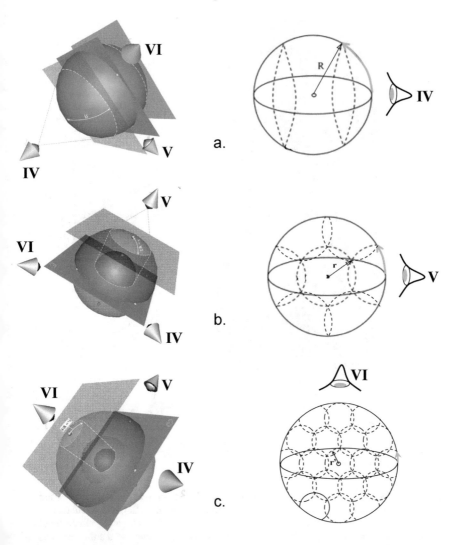

Fig. 2.41 Location of independent observers IV, V and VI with respect to three pairs of nodal points of the $_0S_2$ fundamental mode for free oscillations of an R-radius sphere with a three-dimensional Euclidean space oscillation relative to a spherical nodal surface of an R-radius

Observer V, who will already consider the sphere relative to the next pair of nodal and bifurcation points belonging to "cusp I", will detect surface torsional and volumetric spheroidal displacements of r radii, while observer VI, relative to "cusp II", will find surface torsional and body spheroidal and radial displacements of the r' radii (Figs. 2.45, 2.46 and 2.47). Thus, surface torsional displacements of R, r and r' radii, localised with respect to three pairs of its nodal and bifurcation points, appear on the surface of the sphere with body spheroidal and radial oscillations of R, r and r'

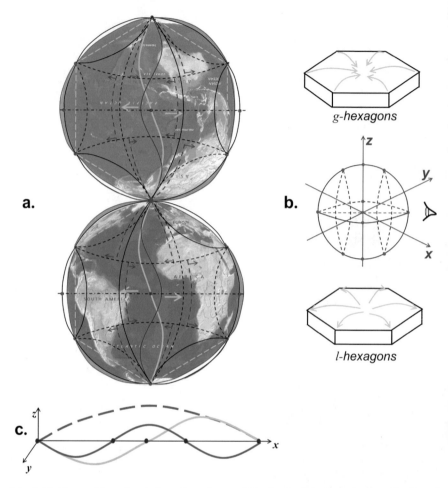

Fig. 2.42 The position of an independent observer IV, who, in the spherical space of radius R, sees the second and first overtones of the sphere's free oscillations in the form of bulk and surface Rayleigh and Love waves (**a**) and the oscillations of individual particles along closed trajectories in the form of g- and l-hexagon dissipative structures (**b**). The second and the first overtones of a vibrating string (**c**)

radii localised with respect to the nodal point at its centre. These oscillations manifest themselves not only in the form of corresponding deformations of its surface, but also lead to localisation of the external and internal core (Figs. 2.45, 2.46 and 2.47).

It turns out that the decomposition into bifurcation modes, which is inherent in the absolute majority of real physical systems, is due, firstly, to their nonlinearity and, secondly, to a sufficiently large number of degrees of freedom. It is worth recalling that, under normal conditions, a cubic centimetre of gas contains about 10^{19} particles interacting with each other; here, it is wave processes and bifurcations that provide the long-wave correlations. From a mathematical point of view, free oscillations of

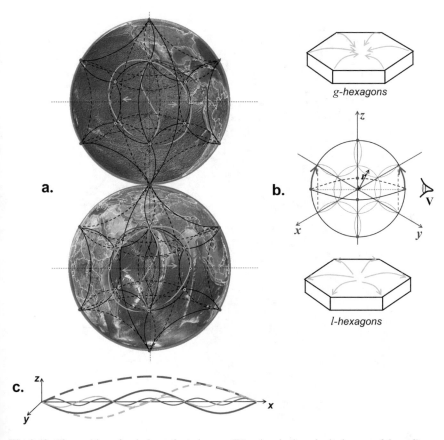

Fig. 2.43 The position of an independent observer IV seeing, in the spherical space of the radius r, the second and the first overtones for free oscillations of the r-radius sphere in the form of bulk and surface Rayleigh and Love waves (**a**), as well as the oscillations of individual particles along closed trajectories in the form of g- and l-hexagon dissipative structures (**b**). Displacements are shown only with respect to one pair of nodal points. The fourth and the third overtones for free oscillations of a string (**c**)

space with respect to special nodal and bifurcation points determine the relationship of discrete oscillations for individual particles, described by ordinary differential equations, with wave processes presenting collective motions in distributed systems described by partial differential or integral equations.

On the other hand, if we look at this problem not only from the standpoint of geometry, but also physics, we can say that free oscillations of space with respect to special nodal and bifurcation points relate, ultimately, only to finite-dimensional (conservative) systems with continuous-time and not-infinite-dimensional space as a whole. As we know, any such finite-dimensional system with continuous time can be associated with some point mapping, thus ensuring the inextricable connection of these systems with Newtonian trajectories.

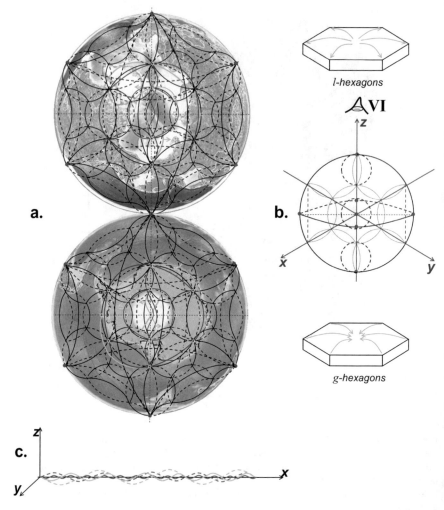

Fig. 2.44 The position of an independent observer who, in a spherical space of radius r', sees the sixth and the fifth overtones for free oscillations of a R-radius sphere in the form of bulk and surface Rayleigh and Love waves (**a**) and discrete oscillations of individual particles along closed trajectories in the form of l- and g-hexagon dissipative structures (**b**). Displacements are shown only relative to one pair of nodal points. The sixth and the fifth overtones for free oscillations of a string (**b**)

One of the main provisions in the fundamental laws of development for these finite-dimensional conservative systems consists of the requirement for the conservation of energy, phase volume and symmetry, as well as their invariance with respect to the Galileo transformation. This term was proposed by Frank (1909) for the description of transition from one inertial reference system (IRS) to another parallel moving inertial reference system (IRS). Using the Galilean transformation formulas, one can

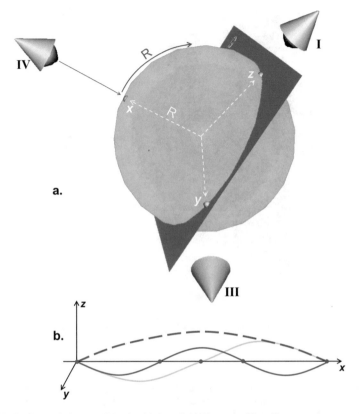

Fig. 2.45 Position of observer IV, who, in the spherical space of R radius, sees the second and the first overtones for free oscillations of a sphere relative to its surface nodal point IV (**a**). The second and the first overtones of a vibrating string (**b**)

easily verify that, if these equations are valid in the coordinate system associated with fixed stars, then they will also be valid in all other reference systems moving linearly and uniformly with respect to these fixed stars. This requirement follows directly from the locality of description for free oscillations of Newtonian systems knowing nothing about the infinite-dimensional structure of atoms and the Universe as a whole.

It can be easily seen that, in this case, we obtained the formulations and solutions of the famous Henri Poincaré theorem on the presence on the surface of an oscillating sphere of "…at least two pairs of fixed points with the position unchanged by oscillation transformations". Henri Poincaré published these topological representations in the form of an article "On a geometric theorem" in 1912 with the following explanations: "Never before have I been in print with such unfinished work… On the other hand, the value of the subject is too great and the totality of the results already obtained is too significant for me to decide of uselessly abandoning them. I can hope that the geometricians who are interested in this problem and, no doubt,

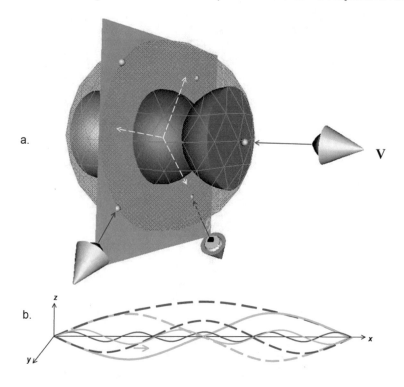

Fig. 2.46 The position of observer V, who, in the spherical space of R radius, sees the forth and the third overtones for free oscillations of a sphere relative to its surface nodal point V (**a**). The fourth and the third overtones for free oscillations of a string (**b**)

happier than me, will be able to derive some benefit from these results and apply them in order to find the right path". In this regard, these "geometric theorems" by Henri Poincaré really turned out to be connected with the fundamental problem of all modern natural sciences—the problem of reference points providing us a physical meaning to all these oscillation and wave phenomena of nature when formulating and solving the wave equation with respect to special nodal and bifurcation reference points.

Having created a graphical algorithm for describing the relationship of the radial, spheroidal and torsional free oscillations of the Earth, we now set ourselves the more ambitious task of solving the wave equation and creating a graphical algorithm for describing the relationship of the periods for the radial, spheroidal and torsional free oscillations of the Earth with the propagation velocity associated with these P- primary and S- secondary bulk and surface seismic waves.

As is known, in 1897, the bulk P- primary and S- secondary waves were singled out by R.D. Oldham from the Geological Survey of India. At that time, they were called the first and second (usually weak) "preliminary oscillations", respectively. The fastest of these waves is the primary or P-wave, under the propagation of which

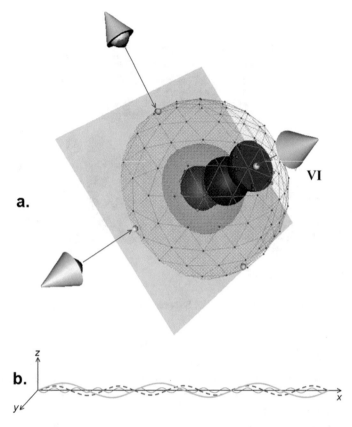

Fig. 2.47 Position of observer V, who, in the spherical space of R radius, sees the third and fourth overtones for free oscillations of the sphere relative to its surface nodal point V (**a**)

elastic rocks undergo compression and tension. An elastic wave of the second type, propagating only in solids, is represented by a secondary or S-wave. This can be imagined as a "concussion", since its passage is associated with the oscillation of rock particles at right angles to the direction of wave propagation, like an vibrating violin string. The nature of the movement is illustrated in Fig. 2.48.

When an S-wave arises in the Earth, elastic rocks undergo shear and torsion strains. Since a liquid cannot be displaced, S-waves cannot pass into the liquid shells of the Earth.

The actual velocity of seismic waves depends on the elastic properties and density of the rocks they passing through. In surface rocks of the granite type, the characteristic value of the P- and S-wave velocity comprises about 5.5 and 3 km/s, respectively. In the deep regions of the Earth where the rocks are compressed, the propagation velocity exceeds 11 and 7 km/s for P- and S-waves, respectively. It turned out that the period of the Earth's radial free oscillations is equal to 20.5 min corresponding to the propagation time of the P- primary and S- secondary body seismic waves from

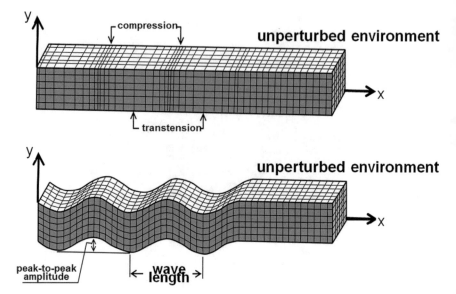

Fig. 2.48 Two types of body, P- primary and S- secondary waves. (Bolt 1984)

the earthquake epicentre to the antipodal point lying on the opposite surface of the Earth (Fig. 2.49). At the same time, the perspective reference point for primary and secondary seismic waves will correspond to the epicentre point of the earthquake near the Earth's surface, and the reference point for the radial free oscillations of the Earth will be in its centre (Fig. 2.49).

By establishing the relationship between the periods of the Earth's radial free oscillations with the propagation velocity for the *P*- primary and *S*- secondary body seismic waves associated with these oscillations, these wave equation solutions directly follow from the irreconcilable contradictions that originally exist between the three "great geometries":

- three-dimensional Euclidean geometry (300 BC) or spherical geometry with a constant curvature of space;
- external, infinite-dimensional Lobachevskian geometry (1826) with varying negative curvature of space;
- internal, infinite, but still finite-dimensional, Riemannian geometry (1854) with changing positive curvature of space.

As shown in Fig. 2.50, the number of lines that can be drawn through a given point parallel to given line is:

- equal to unity in three-dimensional Euclidean geometry or geometry with constant space curvature;
- equal to infinity in external infinite-dimensional Lobachevsky geometry with varying negative curvature of space;

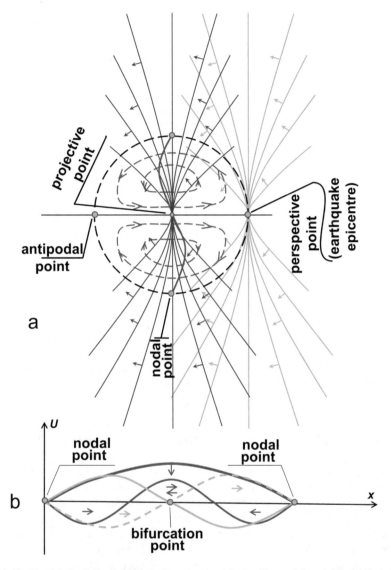

Fig. 2.49 The relationship of the three "great geometries" with special nodal and bifurcation, perspective and projective reference points for the oscillating Earth (**a**); special nodal and bifurcation points of origin in a vibrating string (**b**)

- equal to zero in internal infinite, but still finite-dimensional, Riemann geometry with a changing positive curvature of space.

When describing natural oscillation phenomena, the irreconcilable a priori contradictions between the three "great geometries" with constant and changing negative and positive curvature of space are not only the cause for the appearance of triaxial

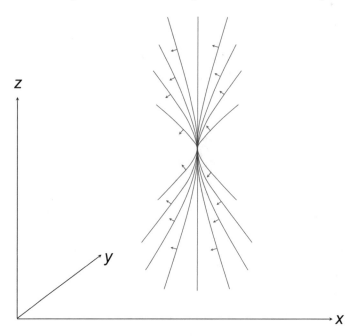

Fig. 2.50 Relationship of the three "great geometries". Explanations in the text.: the number of lines that can be drawn through a given point parallel to given line is: equal to unity in three-dimensional Euclidean; equal to infinity in external infinite-dimensional Lobachevsky geometry; equal to zero in internal infinite, but finite-dimensional Riemann geometry

deformation ellipsoids in the classical theory of elasticity (Fig. 2.51) and the decomposition of the sphere's free oscillations in spherical harmonics (Fig. 2.52), but also for the motion of material particles along closed Newtonian trajectories (Fig. 2.53).

Already at the first seismograms of 1897, in addition to radial or "preliminary oscillations" with the periods related to the propagation velocity of the radiant energy of the P-primary and S-secondary waves through the Earth's body, "big waves" were also identified by R. Oldham and later named after the famous English mathematicians A. E. Love and Lord Rayleigh as the waves of Love and Rayleigh.

The spheroidal and torsional free oscillations of the Earth associated with the propagation velocity of these later-arriving waves had significant amplitudes and durations. It was soon established that they run along the surface of the Earth. For example, they can be compared with waves moving directed by an outer surface, like the sound waves propagating around a "whispering gallery" in the dome of St. Paul's Cathedral in London.

When superimposed, the Love L- surface waves propagating along the Earth's surface produce T torsional modes with a fundamental oscillation period $_0T_2$ equal to 44 min, which corresponds to the time of their passage from the epicentre of an earthquake to the antipodal point on the Earth's surface (Fig. 2.56).

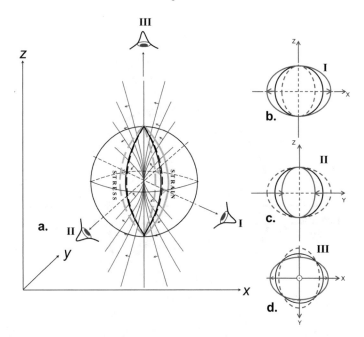

Fig. 2.51 Irreconcilable contradictions between the three "great geometries" with constant and changing negative and positive curvature of space as the reason for the appearance of triaxial deformation ellipsoids in the classical theory of elasticity

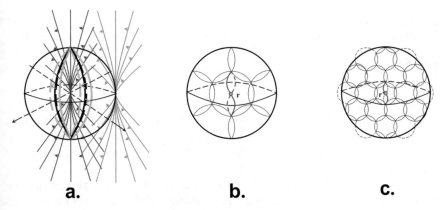

Fig. 2.52 Irreconcilable contradictions between the three "great geometries" with constant and changing negative and positive curvature of space as the reason for the appearance of triaxial ellipsoids of decomposing free oscillations of the sphere by spherical harmonics

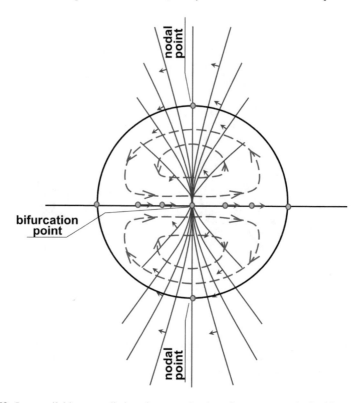

Fig. 2.53 Irreconcilable contradictions between the three "great geometries" with constant and changing negative and positive curvature of space as the cause for the movement of matter particles along closed Newtonian trajectories in the phenomena of thermogravitational and gravitational instability

The second type of surface waves—Rayleigh waves—R, named after the English mathematician Lord Rayleigh, add up to form S spheroidal modes with a fundamental oscillation period $_0S_2$ equal to 54 min, which also corresponds to the transit time from the earthquake epicentre to the antipodal point on opposite surface of the Earth (Fig. 2.56).

These solutions of the wave equation, which establish the relationship between the periods of the spheroidal and torsional free oscillations of the Earth with the velocity of propagation for the surface waves of Love and Rayleigh, also follow from the a priori irreconcilable contradictions between the three "great geometries". Indeed, if we draw a triangle on a spherical surface with a constant and unchanging curvature of space within the space–time model of the relationship between the three "great geometries" shown in Fig. 2.57, we find that the sum of the angles in this triangle is:

– equal to two right angles in the geometry of Euclidean or geometry with a constant curvature of space;

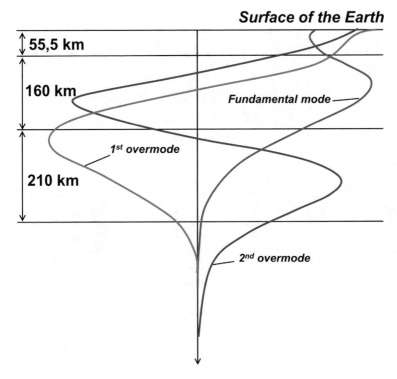

Fig. 2.54 The distribution pattern for in-depth displacements in the bowels of the Earth for three modes of Love waves with a period of 30 s (Bolt 1984)

– less than two right angles in the Lobachevsky geometry with varying negative space curvature;
– more than two right angles in the Riemann geometry with varying positive curvature of space.

The same can be said about the ratio of the areas of these triangles. In this case, any deformation of the space will lead to the propagation of surface waves and the appearance of spheroidal and torsional free oscillations at all scales of organisation in the Universe.

This relationship of the propagation velocity for P-primary and S-secondary bulk and surface seismic waves of Love and Rayleigh with periods of radial, spheroidal and torsional free oscillations of the Earth was the basis for solving the wave equation and creating a new measurement system that, according to the figurative expression of Albert Einstein, "…connects the processes taking place now and here with the processes that will take place a little later and at some distance".

The illustrative spatial–temporal model defining the relationship between the periods of the Earth's radial, spheroidal and torsional free oscillations with the propagation velocity of bulk and surface seismic waves relative to special nodal and bifurcation, perspective and projective reference points within the framework

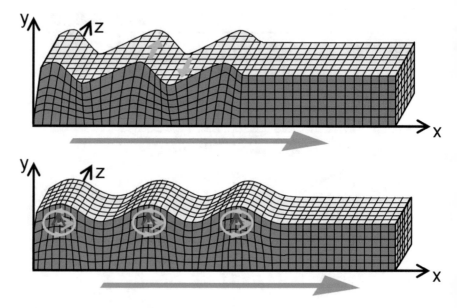

Fig. 2.55 Soil displacements on the Earth's surface during the passage of Love (**a**) and Rayleigh (**b**) waves (Bolt 1984)

of three "great geometries" is shown in Figs. 2.49 and 2.56. It is evident that, as in the case of a vibrating string, when, on the one hand, it is considered as an infinite set of oscillating discrete and tending to limit masses pendulums interconnected by weightless threads (or springs) and, on the other hand, in the form of propagating with the limiting velocity of continuous travelling waves not related to the limiting masses, we can also obtain new limit relativistic solutions of the wave equation in the framework of this model.

As in the case of a vibrating string, in the transition to a new visible event horizon associated with the velocity of light propagation, the spatiotemporal coordinates for a perspective reference point are determined here by the famous Einstein equation and are connected in the framework of the special theory of relativity with the Lorentz space–time transformations. Moreover, the irreconcilable contradictions originally existing between the three "great geometries" determine the discrete-wave nature of the propagation for travelling waves and light. This can be easily seen by looking at the propagation model of seismic travelling waves relative to the perspective reference point within the framework of three "great geometries" with both a constant and a variable velocity reaching the velocity of light propagation by negative and positive curvature of space (Fig. 2.49). This model simultaneously describes the propagation of wave energy not only in the form of rectilinear rays and wave fronts, but also in the form of discrete massless pulses with the dimensionality determined by the

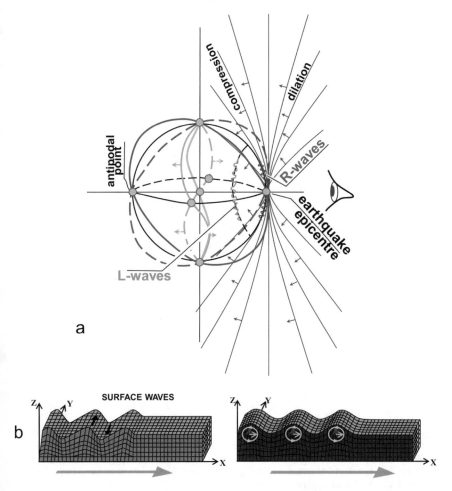

Fig. 2.56 Interrelation of *L*- Love (**b**) and *R*- Rayleigh (**c**) surface seismic waves with *S*- spheroidal and *T*- torsional free oscillations of the Earth. Explanations in the text.: Love *L*- surface waves produce *T* torsional modes with a fundamental oscillation period $_0T_2$; Rayleigh waves *R* add up to form *S* spheroidal modes with a fundamental oscillation period $_0S_2$

velocity of surface wave propagation and the frequency of spheroidal and torsional oscillations. The presence of these discrete spheroidal and torsional waves associated with surface waves explains not only the laws of reflection for travelling waves when they, while maintaining the angle of incidence, like billiard balls, bounce off the reflecting surface, but also the presence of these oscillation pulses of rotational spin, which directly follows from the a priori irreconcilable contradictions between the three "great geometries". It is not difficult to guess that, from the topological point of view, in this case, the existence of surface waves and related spheroidal and torsional momenta represents visual formulations and solutions of the famous Henri Poincaré "two-dimensional sphere" theorem.

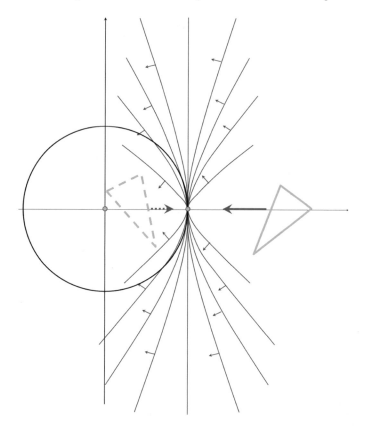

Fig. 2.57 Relationship of the three "great geometries" with the sum of the triangle angles and its area. If we draw a triangle on a spherical surface with a constant and unchanging curvature of space, we find that the sum of the angles in this triangle is: equal to two right angles in the geometry of Euclidean or geometry; less than two right angles in the Lobachevsky geometry; more than two right angles in the Riemann geometry

In order to clarify this difficult issue, at the beginning of the twentieth century, Einstein conducted a deep critical analysis of experimental methods for measuring space and time. At the same time, he adopted the postulate, according to which no energy and signal can propagate at a velocity exceeding the velocity of light in a vacuum; moreover, that the propagation speed of light in a vacuum is constant and does not depend on the direction of its propagation. The existence of this upper bound for signal propagation velocity made it possible to derive the Lorentz transformation formulas and to understand their physical meaning.

As a result of this, the speed of light has acquired the status of a fundamental physical constant. Time turned out to be inextricably linked with space. These new spatiotemporal characteristics of the world around us determined the geometric and physical representations of the special theory of relativity.

When describing free oscillations of a string and a sphere "within the framework of three "great geometries" under a constant and variable with a velocity, not exceeding the velocity of light propagation, negative and positive curvature of the space, the group structure of these transformations provides us with invariance of all fundamental laws related to the velocity of wave and light propagation relative to the spatiotemporal transformations of Lorentz:

- transformation of the gradient reflecting the spatial variability velocity of the tensor components;
- divergence transformations or a transformation reflecting the presence of a solenoidal field;
- transformation of the rotor representing the existence of vortices in nature.

These transformations of coordinate systems in a perspective metrics provide us not only with a derivation of the Lorentz transform formulas, but also with a better understanding of their geometric and physical meaning within the framework of the three "great geometries". Confirming the inextricable connection between geometry and physics, the perspective measuring system, when combined with Einstein's special theory of relativity, answers the question: why did the velocity of light not only began to determine the geometric features of the new coordinate system, but also acquired the status of a fundamental physical constant; why time turned out to be inextricably linked with space, but quantum mechanics – with its changing negative and positive curvature? Thus these new spatiotemporal characteristics of the world around us can be seen to determine the geometric and physical representations of the special theory of relativity:

- the velocity of light does not depend on the movement of the source (since, in the perspective metrics, the coordinate system is determined by the velocity of light and this velocity does not change when moving from one inertial system to another);
- the relativity of simultaneity in inertial reference systems moving relative to each other (the result of relative motion in perspective metrics and its connection with the velocity of light evolves Lorentz's time dilation and contracting or broadening).

Perspective metrics, like the special theory of relativity, dealing with the radiation energy of waves and wave processes simultaneously unfolding in time and space, defines a new "distance" introducing a new invariant. This is the "distance" between two spatiotemporal events relative to a perspective point in space. Observers moving uniformly and rectilinearly relative to each other here attribute different numbers to the spatial distance between two points or the time elapsed between two events. However, they attribute the same value to the space–time interval associated with the velocity of light propagation that separates these two events. According to the special theory of relativity, the time interval between two events corresponding to a ray of light propagating in a void is zero. As a result, the velocity of light has the same value for all observers in a perspective metrics, i.e. is a universal constant.

Since that time, Einstein's principle of relativity has been adopted as a new formulation of the fundamental laws of nature, referring to the invariance of all physical laws related to the velocity of light propagation interpreted in isolation from the dual physical properties of oscillation and wave processes, as well as from the general system of independent observers capable of describing the total energy as progressing and standing waves. This misunderstanding of the topology dual ideas for travelling waves is reflected in the logic adopted by Albert Einstein to construct the theory of relativity with new properties of motion at high velocities derived from the relativistic properties of space and time. From the position of topology, this is only an extreme and particular case of the special theory of relativity for all waves progressing relative to the velocity of light propagation.

Subsequently, Einstein himself pointed out this drawback of the theory he had constructed, noting in his creative autobiography the inappropriateness of separating scales and clocks from the rest of the world of physical phenomena. "You can see," he wrote, "that the theory introduces (in addition to four-dimensional space) two kinds of physical objects... This, in a sense, is illogical: in fact, the theory of scales and clocks should be derived from the solutions of the basic equations and not be considered independent from them" (Einstein 1923).

In fact, as our experience in describing free oscillations of a string and the Earth showed, the difficulties of the relativity theory consisted not in the legitimacy of using the Lorentz transformations, but in a misunderstanding of the fact that, in order to describe the total energy of oscillation and wave phenomena of nature, it was necessary to further expand the relativistic principles underlying spatiotemporal transformations of fundamental laws regarding new visible event horizons related to the velocity of light propagation not only for travelling, but also for standing waves associated with discrete masses.

Indeed, by solving the dual wave equation for progressing and standing waves in the framework of three "great geometries" during the transition to new visible event horizons related to the velocity of light propagation, we will deal simultaneously not only with perspective reference points for travelling waves having a propagation velocity that cannot exceed the velocity of light, but also projective reference points for standing waves having spatiotemporal properties associated with any physical body having a mass whose escape velocity is equal to the velocity of light propagation at its surface, both from the position of classical Newtonian physics and general theory of relativity. And as result, these projective points become invisible to the observer ("black holes"). These "massive" objects can be electrically charged or neutral, with an extremely high internal temperature or be absolutely cold, can represent a black hole of the first or second kind and can possess spin.

Thus, solving the wave equation for standing waves as in the case of a vibrating string in the framework of relativistic physics, we first concentrate the entire mass of the oscillating sphere at one point corresponding to its centre of gravity and the dimensions of its gravitational radius, i.e. becoming a projective point of space invisible to an observer with an existing visible event horizon associated with the velocity of light propagation. Next, we expand the of the string mass in the form of an infinite set of oscillating discrete pendulums tending to the maximum Planck sizes and masses that provide, through Poincaré resonances, fundamental gravitational interactions, which also lie beyond the boundaries of visible event horizons and decrease, according to the current law of universal gravitation, in proportion to the square of the distance from the centre of mass concentration.

Arising in the framework of the "three great geometries" due to resonances, the fundamental gravitational interactions between the Planck masses require us to quantise the geometry of space–time itself. The Planck time of the order of 10^{-44} s will correspond here to the time the light travels a distance equal to the Planck length of the order of 10^{-33} cm. On the other hand, this time will correspond to the period of free oscillations for the discrete Planck masses. Moreover, the right-hand side of the Einstein equation—the energy momentum tensor of matter—becomes a quantum operator, which, in this case, evolves a relativistic generalisation of the energy and momentum concepts of classical continuum mechanics.

These solutions of the wave equation for standing waves, connected in the framework of three "great geometries" with free oscillations of discrete pendulums tending to the Planck limit mass, length and time and representing in this sense the solutions of the famous Henri Poincaré's theorem for the three-dimensional sphere as an everywhere-dense "gapless" set, were called by us the spatiotemporal Poincaré transformations.

Thus, within the framework of three "great geometries" reflecting the features of the three-dimensional Euclidean perspective and projective perception of the world around us, having created a graphic algorithm for describing the relationship of the periods for the Earth's radial, spheroidal and torsional oscillations with the propagation velocity of primary and secondary bulk and surface seismic waves associated with these oscillations and, further, having moved on to new visible—and, more precisely, invisible—event horizons covering dimensionalities from Planck values and Schwarzschild radii to the velocity of light propagation in the Universe, we came to a unified theory of space–time transformations of the fundamental nature laws of Galileo, Lorentz and Poincaré by presenting them in the form of three "great theorems".

2.3   Formulations and Solutions of the Fundamental Wave Equation of Nature

Einstein's dream of a unified theory that would include all interactions remains alive today.

Steven Weinberg.

Steven Weinberg.
Laureate, Nobel Physics Prize, 1979.

The new formulations of the fundamental laws of nature, striving to describe all interactions in the form of a single theory, as Albert Einstein rightly remarked, owe their origin to "… not the rise of fantasy, but the irresistible force of experimental facts …", which, when studying the characteristics of the wave and light propagation, led to the creation of three "great geometries":

– three-dimensional Euclidean geometry or spherical geometry with constant space curvature;
– infinite-dimensional Lobachevskian geometry (1826) with varying negative space curvature;
– infinite, but still finite-dimensional, Riemannian (1854) geometry with a changing positive space curvature.

It turned out that a priori irreconcilable contradictions and unexpected relationships existed between these three "great geometries", referred to in terms of duality, which determined the fractal (or fractional) structure of the space surrounding us. The dualities show that the three "great geometries" describe various aspects of the same physical reality, which in no case could be opposed to each other. We have presented these a priori irreconcilable contradictions and unexpected relationships between the three "great geometries" within the single fractal (or fractional) geometry of Nature in the form of a simple and intuitive graphic model. The conclusions that can be drawn from this model, which are very important for the whole of modern natural science, reflect the experimentally-established features of wave and light propagation (Fig. 2.58).

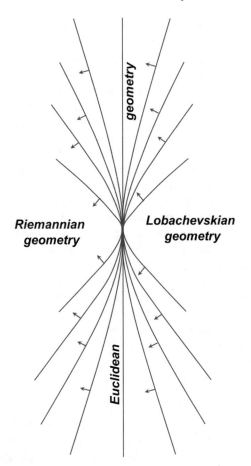

Fig. 2.58 The relationship of the three "great geometries". The initially existing irreparable opposites and unexpected relationships between the three "great geometries" within the framework of
single fractal (or fractional) geometry of Nature are presented in the form of simple and visual
graphic model, from which one derives the conclusions that reflect the experimentally established
features of the propagation of waves and light

Within the framework of these three "great geometries", it is easy to see that the
space around us acquires a fractal structure as a result of a simultaneous change
in negative curvature in the infinite-dimensional Lobachevsky geometry and positive curvature in the infinite, but still finite-dimensional Riemannian geometry with
respect to the straight line separating these two spatial regions belonging to the
Euclidean geometry (Fig. 2.58). In this fractal space, the basic properties of wave
propagation in the form of rectilinear rays and wave fronts are described, including
their remarkable properties of reflection from the surface, like discrete billiard balls,
when the angle of their incidence on the reflecting surface is equal to the angle
of reflection, their dispersion, interference, diffraction and geometric discrepancy.

The last property of wave propagation, which is associated with a decrease in their amplitude and is inversely proportional to the squared distance from the perturbation centre, occurs simply because the wave energy is distributed in an increasingly vast space. Awareness of the fact that the features of wave propagation described within the three "great geometries" are the basis of a unified system of mensuration, of course, is a very significant achievement of all modern natural science.

On the other hand, in our case, among the most important discovery of modern science, expanding the foundations of not only Newtonian physics and thermodynamics, but also of the general and special theories of relativity and contemporary string theory, are the experimental data on the relationship between the propagation velocity of bulk and surface seismic waves from the earthquake epicentre to the antipodal point lying on the opposite surface of the Earth, with the periods of radial, torsional and spheroidal free oscillations of the Earth interconnected with these seismic waves (Fig. 2.59). Along with special nodal and bifurcation reference points, it turned out that there are additional perspective and projective reference points characterising the relationship of oscillations and waves within Newtonian physics, in whose vicinity the geometric and dynamic characteristics of space are determined by the rate of change of negative and positive space curvature with respect to a straight line separating these regions belonging to Euclidean geometry (Fig. 2.59).

The relationship between the special nodal and bifurcation, perspective and projective reference points, which are needed to describe the free oscillations of the Earth or sphere within the three "great geometries", is presented in analogy to the free oscillations of a string Fig. 2.59 in the form of a simple and clear graphic model.

During the transition to new event horizons related to the speed of light propagation, gravitational radii and Planck quantities, this graphic algorithm opens up the possibility for solving the wave equation for the free oscillations of our entire Universe, promising to create a unified theory of nature that describes the total energy of vibrational and wave phenomena. We have presented these formulations and solutions of the fundamental wave equation of nature within the three "great geometries" in the form of three "great theorems", i.e. by Galilean, Lorentzian and Poincaréan space–time transformations of the fundamental laws of nature with respect to special nodal and bifurcation, perspective and projective reference points.

Within this unified theory of nature, determined by the space–time transformations of Galileo, Lorentz and Poincaré, the moment of the birth of our Universe certainly belongs to one of the most unusual natural phenomena ever studied by science. Since it doesn't have direct analogues in any of the other areas of modern science, it comprises in this sense a "moment of truth" to test the basic provisions of the unified theory of Nature.

In historical terms, first attempts to formulate and solve the fundamental wave equation can be associated with the date of November 27, 1783, when John Michell (1724–1793), a clergyman from the village of Thornhill in Yorkshire, England—who has also been called the father of seismology—sent a letter to the Royal Society in which Newtonian celestial mechanics and corpuscular optics were for the first time reconciled. The letter referred to the concept of a massive body in which gravitational

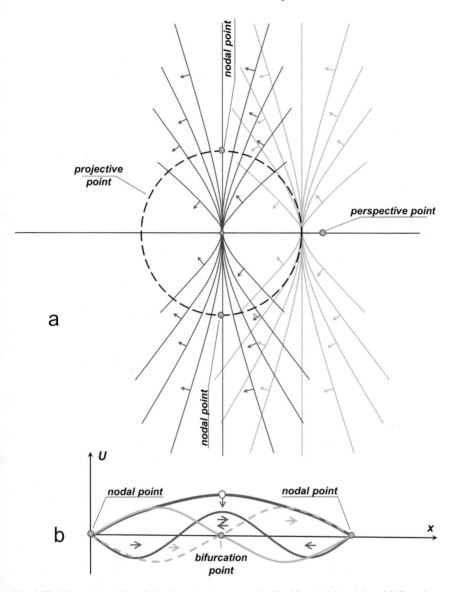

Fig. 2.59 The relationship of the three "great geometries" with special nodal and bifurcation, perspective and projective reference points for the oscillating sphere (**a**) and the vibrating string (**b**)

attraction was so great that the speed necessary to overcome this attraction (escape velocity) was equal to or greater than the speed of light. In accordance with the motto of the Royal Society—"*Nullius in verba*" (from Latin "take nobody's word for it", meaning that the proof should consist in experiments and calculations, but not the words of any authorities), a calculation was included in the letter from which it followed that for a body having the same density as the Sun but a radius of 500 solar radii, the escape velocity on its surface would be equal to the speed of light. Thus, since it would be impossible for light to leave such a body, as a consequence it would be invisible to the observer. The equatorial radius of such a body would be equal to $\sim 3.48 \cdot 10^{11}$ m, a volume of $\sim 1.75 \cdot 10^{35}$ m^3, a mass of $\sim 2.47 \cdot 10^{38}$ kg, an average density (in accordance with the calculation conditions) of ~ 1.409 g/cm^3 and an escape velocity on the surface of $\sim 3 \cdot 10^8$ m/s.

On the other hand, within the general theory of relativity, the gravitational radius, on the surface of which the escape velocity would be equal to the speed of light propagation according to the coordinates of Riemannian geometry was first calculated in 1916 by the German astronomer and physicist Karl Schwarzschild. This gravitational radius was found to be proportional to the mass M of the body and expressed by $r_g = 2GM / c^2$ where G is the gravitational constant and c is the speed of light in a vacuum. This turned out to be the only spherically symmetrical exact solution of the Einstein's equation in empty space. In particular, this Schwarzschild metric describes quite accurately the gravitational field of a solitary non-rotating and uncharged black hole, as well as that of a solitary spherically-symmetric massive body.

The gravitational radii of ordinary astrophysical objects are negligible as compared to their actual sizes: for example, for the Earth $r_g \approx 0.887$ cm, while for the Sun $r_g \approx 2.95$ km. At the same time, the gravitational radius of an object with the mass of the observable Universe would be approximately 10 billion light years (Deza 2012), which makes the effects of Newtonian physics and the general theory of relativity important in studying such objects to create a theory of the birth of our Universe.

According to its properties, the Schwarzschild gravitational radius calculated for the Sun $r_g \approx 2GM / c \approx 2.95$ km coincides with the radius of a spherically symmetric body $R_G \approx 3.48 \cdot 10^8$ km of 500 solar radii and the density of the Sun calculated by John Michell, because in classical mechanics the escape velocity on their surface is equal to the speed of light $\vartheta_2 \approx \sqrt{2GM/R} \approx 300\ 000$ km/s. The fundamental nature of the relationship of these radii with the dimensions of the Sun within the three "great geometries" is shown in Fig. 2.60.

This fact, of course, is not accidental, but a consequence of the fact that classical mechanics and the Newtonian theory of gravity in the transition to a new event horizon associated with the speed of light propagation are contained in the general theory of relativity as a limiting case (Ginzburg 1980). But even more surprising is the fact that such an interconnection between the general and special theory of relativity of Albert Einstein and Newtonian physics within the three "great geometries" opens up the possibility for us to solve the wave equation for the free oscillations of our Universe by analogy with the wave equation for the free oscillations of the Earth and free oscillations of the string in the form of standing and traveling waves, taking into account the new visible event horizons associated with the speed of light propagation, gravitational radii and Planck values.

In this case, we will not be interested in surface waves that run relative to the perspective reference point, the propagation velocity of which cannot exceed the

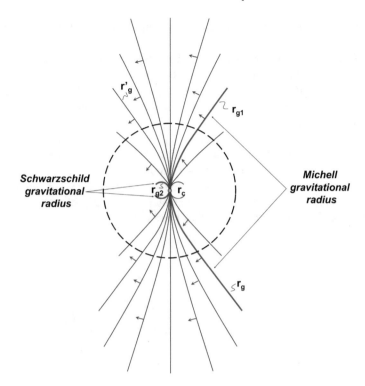

Fig. 2.60 Fundamental relationship of the mass and size of the Sun with the gravitational radii of Michell and Schwarzschild within the three "great geometries". For explanations see the text

speed of light propagation resulting in their attachment to the surface of bodies with a gravitational radius, but rather in standing internal gravitational waves.

In solving the wave equation for our entire Universe within the three "great geometries" and strictly following the ideas of d'Alembert, we draw back the veil of secrecy of its origin. According to Big Bang theory, the early Universe took the form of a hot plasma consisting of electrons, baryons and constantly emitting-, absorbed- and re-emitting photons. Thus, the radiation was in a state of thermal equilibrium with matter, with its spectrum corresponding to the spectrum of a black body.

The existence of this radiation was predicted back in 1948 by George Gamow and his students Ralph Alpher and Robert Herman, who believed that if the Universe in the past was hot and denser than in our time, then it would be "opaque", since photons would have enough energy to interact strongly with matter. They further showed that the fundamental equilibrium between matter and light is disturbed at a temperature of approximately 3000 K; therefore, our Universe only became "transparent" when radiation became "cut off" from substance. In this case, the temperature of this relict radiation in our time should be about 3 K. This epochal prediction anticipated one of the greatest experimental discoveries of the last century, i.e. that of the now famous relict radiation at a temperature of 2.7 K by Arno Penzias and Robert Herman in 1956.

The observed sphere corresponding to this moment, referred to as the Surface of Last Scattering (Klimushkin and Grablevsky 2001), is the most distant object that can be observed in the electromagnetic spectrum. On the other hand, the radius of its observed portion corresponds to the Schwarzschild radius or the gravitational radius of our entire Universe (Fig. 2.61).

The gravitational radius of the observable part of our Universe is $r_g \approx 2GM_x / c^2 \approx 4.6 \cdot 10^{10}$ light years; where G is the gravitational constant, c is the speed of light in vacuum, M_x is the characteristic mass of more than 10^{53} kg (Linde, ...). Thus, the Universe, observed from the centre of the Solar system (the observer's place of residence), comprises a sphere having a diameter of about 93 billion light-years and a volume of $3.5 \cdot 10^{80}$ m^3 or $8.2 \cdot 10^{180}$ Planck volumes.

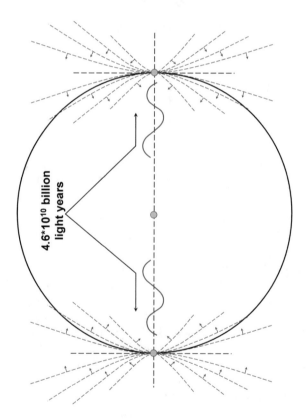

Fig. 2.61 The Universe observed simultaneously in perspective and projective mensuration. On the one hand, for an observer who at its centre, this consists of the most distant object that can be observed in the electromagnetic spectrum. On the other hand, the radius of its observed part corresponds to the gravitational radius of the observed part of the Universe. Further explanations are in the text

This is the relativistic solution of the wave equation for the traveling and standing waves of our Universe. In full accordance with contemporary ideas about its observable parameters, as in the case of a vibrating string, we first concentrate the entire mass of the observable Universe at one point of a size of about 93 billion light-years, which in projective mensuration defines its centre of gravity corresponding to the size of its gravitational radius, i.e. the point on the surface of which escape velocity is equal to the speed of light propagation and it becomes an invisible projective point of space for the observer. Next, we expand its mass in the form of an infinite set of striving for ultimate Planck dimensions $r_g \approx 2GM / c^2 \approx 10^{-35}$ cm and Planck masses $M \approx 2.176 \cdot 10^{-8}$ kg, which also contain a visible event horizon associated with the propagation speed of light $\vartheta_2 \approx \sqrt{2GM/R} \approx 300\ 000$ km/s of oscillating discrete pendulums. This, as we have already noted, is the relativistic solution of the d'Alembert wave equation for standing waves, which ensures, on the scale of Planck time ~ 10^{-44} s and through gravitational Poincaré resonances, fundamental gravitational interactions lying beyond the boundary of visible event horizons. Since the amplitudes of the free oscillations of invisible discrete Planck masses, interconnected by Poincaré resonances, will decrease in proportion to the square of the distance from the centre of their concentration, this, as we noted earlier, will be a relativistic or quantum-wave representation in the form of standing gravitational waves of the fundamental law of universal gravitation discovered by Newton in 1667. In this solution of the wave equation—on the one hand, in the form of discrete pendulums, which are the maximum possible up to a dimensionless coefficient of the order of mass unit, in the mass spectrum of elementary particles invisible to the observer from the point of view of new event horizons related to the speed of light propagation, gravitational radii and Planck values, and, on the other hand, in the form of standing internal gravitational waves—the oscillation frequency corresponds to the Planck unit of time ~ 10^{-44} s, i.e. the boundary between matter and standing gravitational waves is completely "erased" here. It turned out that, in contrast to the special nodal and bifurcation points of Newtonian dynamics, perspective and projective reference points possess affine properties of self-similar sets, that is, in particular, they have no common reference point. Therefore, they will not be found in the centre of the Solar system, i.e. in the location of the perspective observer. These are mutually univocal reflections of one point to another, which form groups in a perspective and projective fractal mensuration systems, related to the perception of the world around us through the speeds and fundamental laws of wave and light propagation, gravitational radii and Planck quantities.

Thus, in solving the fundamental wave equation of nature, we have presented a unified theory of the Universe within the three "great geometries" in the form of three "great theorems" or Galilean, Lorentzian and Poincaréan space–time transformations with respect to the fractal dimension of special nodal, bifurcation, perspective and projective reference points.

This unified theory, which was only intuitively outlined at the beginning of the twentieth century in the works of Henri Poincaré and Albert Einstein, taking the Galilean, Lorentzian and Poincaréan space–time transformations of the fundamental laws of nature into account as control parameters to which all other degrees of freedom of the Universe surrounding us are adjusted, had already by the second half of the twentieth century become the basis of synergetics, offering a new inter-disciplinary approach to the description of complex natural phenomena from the most diverse areas of natural science. It turned out that next to the world of closed Newtonian (or Hamiltonian) systems, there was a completely different world of open dissipative systems, which Henri Poincaré and Albert Einstein were able to see for the first time. Unlike Newtonian or Hamiltonian systems, the phase volume of dissipative systems changes with time and, along with the three-dimensional Euclidean geom-etry, we must simultaneously use the Lobachevsky geometry with varying negative space curvature and Riemannian geometry with varying positive space curvature. In accordance with this, it turned out that the phase space of dissipative systems could contain not only special nodal and bifurcation reference points characterising oscillatory and wave processes in the Newtonian systems, but also new reference points, such as attractors and repellors, which also turned out to be connected with perspective and projective features of the perception of the world around us seen through human eyes.

The attractors are characterised by the fact that all phase trajectories from a certain region of the phase space, called the region of attraction, are asymptotically attracted to them. During the evolution of a dynamical system with an attractor, the volume of a phase droplet decreases indefinitely, i.e. the droplet shrinks to the attractor. However, the attractor itself, having zero measure in the phase space, can turn out to be a nontrivial set, on which motion is stochastic. Moreover, the attractor itself, on which stochastic dynamics is realised, is called a stochastic or strange attractor. The latter term was proposed by David Ruelle and Floris Takens.

A strange attractor, occupying the region of the phase space of zero measure, cannot, however, lie entirely in the plane (since the phase trajectories do not intersect). From a geometric point of view, it is, as a rule, a fractal set characterised by a fractal dimension and is a fractional number.

For repellors, on the contrary, instability is characteristic, i.e. the departure of any trajectory starting in some area of this reference point.

Despite attractors and repellors now being identified in almost all areas of natural science, up until now no one has ever managed to link them with perspective and projective reference points or the irreconcilable a priori contradictions between the three "great geometries of nature". In this sense, the three "great geometries", reflecting the inextricable relationship between the discrete and the continuous in nature, clearly demonstrate the properties of self-similar fractal sets, whose exis-tence was first theoretically established by Mandelbrot (1977). Such sets within irreconcilable contradictions that initially exist between the three "great geometries" are similar to the perfect, everywhere-dense Cantor set of points in mathematics that fills the entire set segment, although the total mass of its points is equal to zero (Fig. 2.58). Within the three "great geometries", these sets become a new system of mensuration and acquire remarkable geometric and dynamic properties that allow the relationship of discrete vibrations and continuous waves to be imagined not only in seismology and geology, Newtonian physics and thermodynamics, but also in phenomena associated with gravity, light and matter in the form of a visual space–time model.

This connection of the micro- and macrocosm within the three "great geometries" topologically manifests itself at all scale levels of the organisation of our Universe in the form of wave-like and cellular structures of gravitational instability. The geometric similarities of these structural forms are revealed, regardless of the nature and scale of their manifestation, even more clearly when we turn to the applied areas of natural science in hydrodynamics, oceanology and meteorology, where among the apparently scattered examples, the same geometric commonality of spontaneous thermogravitational and gravitational structuring is revealed.

We recall similar wave-like and cellular structures of thermogravitational instability discovered and studied in thermodynamics by Bénard (1901) and referred to as Bénard convective cells. But even in the era of the sailing fleet, when colder currents were blocked by warmer ones, sailors often observed similar structures in the ocean, which were called "internal standing gravitational waves" in all textbooks on hydrodynamics.

Major results of the theory of cellular standing internal gravity waves were obtained by George Stokes as far back as 1847 (Stokes 1847). From a physical standpoint, cellular standing internal gravity waves are the oscillations of a density-unstable medium in a gravitational field. If the volumetric element of such a medium is removed from equilibrium, for example, as a result of a mechanical action or by means of thermogravitational instability due to heating from below, the buoyant force will cause its movement opposite to the equilibrium position. During such a vibrational movement, the stratification of the medium in the gravitational field generates a specific natural phenomenon, which is referred to in fluid dynamics as cellular standing internal gravity waves. Along with those of convective heat and mass transfer, these phenomena are connected with the fact that cold layers, which are density-unstable, descend downward under the gravitational force displacing upwards the lighter layers located beneath. In such cases where a medium is density-unstable, its entire volume, like its surface, breaks down into linear or cellular structures, i.e. standing internal gravity waves that are spatially fixed by wave-like displacement lines and characterised by a certain amplitude and wavelength of the wave process.

Similar structural forms of thermogravitational and gravitational instability were also revealed when studying the surface and deep zones of our planet (Fig. 2.62).

The wide range of methodological approaches aimed at revealing such structures includes: seismic tomography data analysis; maps of geoid global anomalies and gravitational potential; graphical and analytical methods of relief decomposition; figures of the Earth's gravitational field; deciphering of space and aerial images; morphometric, structural-geomorphological and structural-geological investigations (Petrov 1992 a, b, 2007, 2019).

The discovery of fractal hierarchies of dissipative structures (cellular standing internal gravity waves of the Earth) is informed both methodically and theoretically by the contemporary state of general knowledge—on the one hand, of thermogravitational (convective) instability in mainly thermodynamic terms, and, on the other hand, of related natural phenomena of cellular standing internal gravity waves, studied independently and in parallel in terms of classic Newtonian mechanics, fluid dynamics

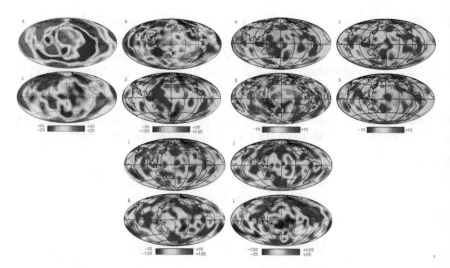

Fig. 2.62 Speed anomaly maps for S362D1, S362C1, and E14 lateral seismic models (Gu et al. 1994; Engdahl et al. 1998) at 12 different depth levels: **a** 100 km, **b** 200 km, **c** 300 km, **d** 550 km, **e** 800 km, **f** 900 km, **g** 1160 km, **h** 1300 km, **i** 1800 km, **j** 2200 km, **k** 2500 km, **l** 2800 km. Double normalisation of the half-tone scale (the scale of the percentage deviation of the lateral velocity component from the average value) refers to the upper and lower maps, respectively

and meteorology. In emphasising the constructive role of dissipative processes in their formation, Ilya Prigogine (1986) referred to such structures as *dissipative*. The examples of dissipative structures in natural and experimental contexts include: Bénard cells (alternation of ascending and descending convective currents in a liquid); plasma striations; the inhomogeneous distribution of concentrations in chemical reactors; Belousov-Zhabotinsky (BZ) oscillating reactions; electromagnetic, acoustic and hydrodynamic waves; cirrus clouds; honeycomb and spiral structures of galaxies; morphogenesis in biological processes; and many other phenomena and processes in physics, chemistry, cosmology, biology, ecology and even sociology.

The empirically-revealed commonality in the description of dissipative structures in various fields of natural science at the beginning of the twentieth century resulted in the creation of synergetics as an interdisciplinary science that studies the paradoxical fact of the self-organisation of natural systems. This science was formed at the junction of the formal sciences and philosophy. However, today, at the beginning of the twenty-first century, a huge empirical basis characterising the generality of the development processes of heterogeneous dissipative structures from an atom or a galaxy to living objects gives us reason to assume the fundamental role of wave processes of matter and space–time transformations of fundamental laws of nature that determine its self-organisation.

Thus, it turns out that both these phenomena—convection and cellular standing internal gravity waves—have a common physical wave nature. Since gravity is a common natural cosmic force to which all substances in motion must conform, such waves can be considered as the most widespread phenomena throughout nature.

Gravitational equilibrium exists in nature only when underlying layers have a higher density than overlying ones. However, this condition is constantly violated both in the Earth's crust and other geospheres, as well as in the hydrosphere and atmosphere surrounding the Earth. In this case, density inversion takes place when the underlying layer is less dense than the overlying one. Such an unstable state is referred to in terms of the Rayleigh–Taylor instability, which leads to the formation of characteristic structural forms of gravitational (convective) instability or cellular standing internal gravity waves of G-, L-hexagon and roll types. All these structural forms and typomorphic structural features have been repeatedly described in terms of natural formations, modelled in laboratories and theoretically studied by many authors, who, however, did not always refer to their common wave nature. Since all wave processes can be deconvoluted into modes, obviously the structural forms formed on their basis must necessarily possess unusual geometric and dynamic properties expressed in the fractional dimension, which, like a set of modes, repeats itself on different scales. On the other hand, by taking this approach, it also becomes possible to solve the inverse problem due to the presence of the fractal dimension and the hierarchy of self-similar structures in natural formations serving as direct proof of the manifestation of the fundamental role of wave properties of matter in formation and structuration processes (Petrov 2007).

Up until now, these studies have been developing in parallel in various fields of natural science, for example, in hydrodynamics and thermodynamics, oceanology and meteorology, geophysics and geology, chemistry and biology, astrophysics and quantum mechanics, but without making the necessary interconnections between each other. Consequently, results pointing to their common underlying wave nature, obtained in one area of natural science, have tended to be little known to researchers working in other areas. Moreover, the question of the manifestation of the wave properties of matter turned out to be based on a very profound insight. Let us recall how nineteenth century scientists were struck by the presence of wave properties of light, which had earlier been imagined as a stream of discrete particles. The concomitant profound revision of the fundamental concepts of physics directly led to the development of quantum mechanics in the twentieth century. It was with good reason that the brilliant French mathematician and physicist Henri Poincaré warned the scientific community in 1893 that the discrete and continuous properties of matter in confluent continuous media cannot be contrasted with each other due to being inextricably linked (Poincaré et al. 1983). A lack of understanding of this fact in modern natural science led to a deep crisis between classical Newtonian dynamics and thermodynamics when describing the similar natural phenomena.

In this book, we have sought to address this gap in contemporary natural science and, on the basis of the manifestation of wave properties of matter in the processes of spontaneous structuring of density-unstable masses in various natural environments, to generalise, at the level of time–space transformations of the fundamental natural laws, the huge quantity of empirical material accumulated by modern geological and geophysical science.

Since the phenomena of the gravitational and thermogravitational instability of the Earth can be associated both with internal sources of heat and matter, as well as with their external sinks (in particular, with the cooling of the Earth through its surface), the Earth is often and fairly compared with a heat engine (Khain 1995, 2003). Our planet can rightly be attributed to the class of open dissipative systems, that is to say, those that can exchange energy, matter and information with the surrounding environment.

The value of dissipative structures for understanding of many natural processes is great. The English scientist Turing (1952) was one of the first to point this out. He decided to investigate one of the most interesting biological processes—morphogenesis—with the aid of mathematical methods.

Turing, who assumed that morphogenesis was governed by chemical processes, was interested in whether it is possible to explain the occurrence of dissipative structures in initially homogeneous tissue based on the simplest chemical concepts. The initial data $\chi_0(x)$, $\gamma_0(x)$ are considered close to spatially homogeneous, but contain small random perturbations λ, which gradually change during development (Turing 1952).

Turing investigated the question how these perturbations develop and what solutions strive for at large times. The results were very interesting. When $\lambda < \lambda_0$, the functions $\chi(x,t)$, $\gamma(y,t)$ strive for a stable, spatially uniform solution. Such a solution is often called a thermodynamic branch. At $\lambda < \lambda_0$, that is, at a certain stage of development, the picture is qualitatively changing. Despite the initial data being close to homogeneous, perturbations increase resulting in the appearance of structure in the medium comprising a spatially non-uniform stationary distribution of concentrations, very similar to the cellular distribution of thermal energy in the experiments of Bénard (1901). At the same time, a second stationary structure appears in the medium, whose output occurs from other initial data (Fig. 2.63).

It is easy to surmise that in the case of morphogenesis and structure formation in geology, we are talking here about internal sources and external sinks of heat and matter (Petrov 2007, 2019). On this basis, in the present work, an attempt has been made to develop a holistic concept of the wave nature of the dissipative processes

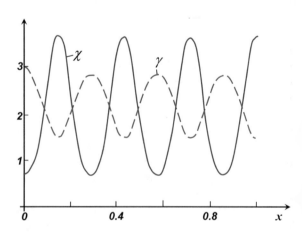

Fig. 2.63 One of the stationary solutions of the Turing problem. This model is referred to as a Brusselator

of the Earth's structure. A new line of research has been developed, whose main provisions as follows:

1. The wave nature of geological processes is the root cause of the structuring and shaping of the Earth. Wave laws allow the connection between heterogeneous density structures to be established and the nature of the motion of matter within the Earth's interior to be determined according to the external topographical structure of the Earth's surface. On this basis, a methodology is developed to allow the distribution pattern of the deep regions of thermogravitational instability to be reconstructed and the motion dynamics of matter in the interior of our planet to be described using the surface forms of the geoid and relief. In this case, the dissipative structures of the Earth become the elementary object of study.

2. Since wave processes and emerging dissipative structures can be represented in the form of an infinite number of modes, they acquire the remarkable geometric and dynamic properties of a fractal or fractional dimension that repeats itself at various scales. In turn, the fractal dimension and hierarchies of self-similar structures can serve as evidence of the manifestation of the fundamental wave properties of matter in the processes of shape and structure formations in nature.

3. The dissipative structures of the Earth are historical categories, since the oscillation frequency of internal standing gravitational waves cannot exceed the buoyancy frequency of the density-unstable geological environment. The oscillating modes of density-unstable processes associated with internal sources and external sinks of heat and matter generate cellular standing internal gravitational waves—dissipative structures of various orders—that allow the causes of the emergence and change of various geodynamic settings in the history of the Earth to be interpreted from new positions.

4. Along with the discrete manifestations of wave processes of structuring of nonequilibrium geological media, the dissipative structures of the Earth determine the hierarchies of geoblock divisibility of the lithosphere at global, supraregional and local levels. As a result, a new approach to the harmonisation of ideas about geological, geomorphological, geophysical, mineralogical and mineral resource components of the Earth's lithosphere is revealed.

Thus, from the standpoint of contemporary natural science, fractal dimensions found in density-unstable geological media—as well as in fractal theory and fractal geometry in general—are considered by us as an abstract mathematical reflection of one of the fundamental wave properties of matter in nature: namely, the deconvolution of oscillatory and wave process into modes simultaneously relative to special nodal, bifurcation, projective and perspective reference points. These fundamental wave properties of matter, which give natural objects such remarkable geometric and dynamic properties, characterise the common development processes of various objects from an atom or galaxy to living beings through Galilean, Lorentz and Poincaré space–time transformations of the fundamental laws of nature.

In this sense, for all structural forms of thermogravitational and gravitational instability, defined as internal standing gravitational waves, the definition of a "blueprint for all the structure in the universe", which was proposed by Stephen Hawking

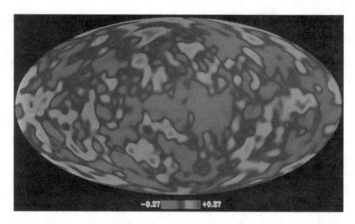

Fig. 2.64 A microwave map of the entire sky obtained by the DMR from Cosmic Background Explorer (COBE), which was launched in 1989. Colour indicates differences in temperature (Hawking 2007)

(Hawking 2007) for a map of the entire sky obtained by the DMR on the COBE satellite, is very appropriate (Fig. 2.64). Without going further into general philosophical and epistemological discussions on this subject, we note the obvious fact that, despite the grandiose theoretical and experimental discoveries of the last two centuries in the form of the dual discrete-wave nature of the structure of matter, as well as the dual nature of standing and traveling waves including light itself, the general and special theory of relativity, the fine atom structure, the torsion spins of elementary particles and their exchange energy, the radiation of black holes and dark matter of the Universe, the tunnel effects of radiation of atoms and elementary particles, and weak and strong fundamental interactions, it would be decidedly impossible to realise the interconnection of the three "great geometries" in the form a unified mensuration system with a unified theory of space–time transformations of the fundamental laws of nature by Galileo, Lorentz and Poincaré without studying the free oscillations of the string and the Earth.

In this sense, the historical significance of our scientific discovery lies in the fact that, with simple and illustrative examples of describing the free oscillations of a string and the Earth, we completed the work begun by Isaac Newton, Jean Lenore d'Alembert, Henri Poincaré and Albert Einstein to create a unified mensuration system for the description of the total energy of vibrational and wave phenomena of nature simultaneously with respect to special nodal and bifurcation, perspective and projective reference points.

Thus, upon transition to new visible event horizons related to the speed of light propagation, gravitational radii and Planck quantities in combining the three "great geometries" with physics, we presented a unified theory of Galilean, Lorentz and Poincaré space–time transformations of the fundamental laws of nature relative to special nodal and bifurcation, perspective and projective reference points in the form of three "great theorems" that underlie the evolution of our Universe.

In this sense, the scientific significance and applied aspects of the unified theory of space–time transformations of the fundamental laws of nature can be correlated with the creation of the theory of the evolution of living nature by Darwin (1859). In both cases, the individual constituent elements of these theories began to form as early as the eighteenth century and were transformed as natural science developed. In wildlife, the origin of biological diversity as a result of evolution was recognised by most biologists during the life of Darwin, however, his theory of natural selection, as the main mechanism of evolution, became generally recognised only in the 50 s of the twentieth century with the advent of a synthetic theory of the evolution of wildlife, almost simultaneously with the formation of a new synergistic approach, studying the commonality of evolutionary processes, both living and non-living nature in all areas of natural science. Today, Darwin's these ideas and discoveries form the foundation of a modern synthetic (ancient Greek $\sigma\upsilon\gamma\theta\varepsilon\sigma\iota\varsigma$ "connection, folding, linking" from $\sigma\upsilon\gamma$- "joint action, complicity" $+ \theta\varepsilon\sigma\iota\varsigma$—"arrangement, placement, distribution") evolution theory that provides an explanation of biodiversity in living nature, just like the ideas and discoveries of a single synergetic (from ancient Greek: $\sigma\upsilon\gamma$, prefix indicating compatibility and $\varepsilon\rho\gamma\upsilon\upsilon$—"activity") mensuration system in the form of three "great geometries" and the unified theory of Galilean, Lorentz and Poincaré spatio-temporal transformations of the fundamental laws of nature are called upon to provide an explanation of the diversity in the development of our entire Universe.

Thus, as a result of formulations and solutions of the fundamental wave equation of nature within the three great Euclidean, Lobachevskian and Riemannian geometries and the three great theorems in the form of Galilean, Lorentz and Poincaré space–time transformations of the fundamental laws of nature with respect to special nodal, bifurcation, perspective and projective reference points, a synergetic picture of the world arises, which is qualitatively different from that presented by classical science.

References

Arnold VI (1990) Theory of catastrophes, 3rd edn. rev. M: Nauka, p 128

Bénard H, Simon LG (1901) Sur les phénylhydrazones du d-glucose et leur multirotation, C.R. Acad. Sci., 4 mars.

Bolt A (1984) Inside the Earth San Francisco. M: Mir, p 189. (in Russ.)

de Broglie L (1965) Revolution in physics: (New Physics and Quanta). In: de Broglie L, Trans. Baklanov SP, Kovrizhnykh LM, Polivanov MK (ed), 2nd edn. M: Atomizdat, p 231. (in Russ.)

Busse FH (1978) Non-linear properties of thermal convection. Res Prog Phys 41: 1931–1965

Darwin C (1859) On the origin of species by means if natural selection, or the preservation of favoured races in the struggle for life. London.

Deza MM, Deza E (2012) Encyclopedia of distances. Springer Science & Business Media, p 644. ISBN 9783642309588

Einstein A (1923) [First published 1923, in English 1967]. Written at Gothenburg. Grundgedanken und Probleme der Relativitatstheorie. [Fundamental Ideas and Problems of the Theory of Relativity] (PDF) (Speech). Lecture delivered to the Nordic Assembly of Naturalists at Gothenburg, 11 July 1923. Nobel Lectures, Physics 1901–1921 (in German and English). Stockholm: Nobel-price.org. (published 3 February 2015)– via Nobel Media AB 2014

Ellis A (1999) Black holes–part 1–history (Archive copy dated 6 October 2017 at Wayback Machine. Astron Soc Edinb J 39

van der Engdahl ER, Hilst RD, Buland RP (1998) Global teleseismic earthquake relocation with improved travel times and procedures for depth determination Bull. seism. Soc Am 88:722–743

Frank P (1909) Sitz. Ber. Akad. Wiss Wien 1909. II a, Bd 218. S373/esp, p 382

Ginzburg VL (1980) About physics and astrophysics. M: Nauka, p 112. (in Russ)

Gu WH, Morgenstern NR, Robertson PK (1994) Post-earthquake deformation analysis of wildlife site. J Geotech Eng 120(2):274–289

Hawking S (2007) The Universe in a nutshell. Trans A Sergeyev. Amphora, Saint Petersburg, p 218. (in Russ.)

Klimushkin DYu, Grablevsky SV (2001) Cosmology. M. (in Russ.)

Khain VE (1995) Core issue of modern geology (geology on the verge of XIX century). M: Nauka, p 190. (in Russ.)

Khain VE (2003) Core issues of modern geology, 2nd edn., revised. M: Nauchny Mir, p 348. (in Russ.)

Lamb H (1882) On the vibrations of an elastic sphere. Proc Lond Math Soc 13: 189.

Levin A (2005) History of black holes. Popular Mechanics. OOO «Fashion Press», №11, p 52–62. (in Russ.)

Mandelbrot BB (1977) Fractals: form, chance and dimension. W.H. Freeman, San Francisco, p 273

Markov MA (2000) Selected works in 2 volumes. In: Matveev VA (ed), vol 1, M: Nauka, p 505, 2001, vol 2, M: Nauka, p 640. (In Russ.)

Petrov OV (1992a) Internal gravity waves of the Earth and non-linear paleogeodynamic dissipative structures. RAS report, vol 326, № 2,p 323–326. (in Russ.)

Petrov OV (1992b) Non-linear phenomena of thermo-gravitational instability and internal gravity waves of the Earth. RAS report, vol 326, № 3, p 506–509. (in Russ.)

Petrov OV (2007) The Earth's dissipative structures: fundamental wave properties of substance. VSEGEI Publishing, Saint Petersburg, p 304. (in Russ.)

Petrov OV (2019) The Earth's dissipative structures. Fundamental wave properties of substance. © Springer Nature Switzerland AG.

Poincaré H (1983) About science. In: Pontryagin LS (ed), M: Nauka, p 560. (in Russ.)

Poincaré H (1974) Selected works in 3 volumes, M: Nauka. (in Russ.)

Stokes GG (1847) On some cases of fluid motion. Philos Mag 31: 136–137

Turing AM (1952) F.R.S. The chemical basis of morphogenesis. Phil Trans R Soc Lond B 237(641), p 37–72

Whitney H (1955) On singularities of mappings of Euclidean Spaces I. Mappings of the plane into the plane. Ann Math 62: 374–410

Conclusion

The scientist's ink and the blood of a martyr are equal in the eyes of the Almighty.

Stephen Hawking. 2001. (c) Photo by Stuart Cohen.

This inscription by the hand of an unknown reader was made on the spread of Louis de Broglie's book "The Revolution in Physics", which is on my library shelf next to the works of d'Alembert, Lobachevsky, Riemann, Poincaré, Einstein, Hilbert, Prigogine, Hawking, Green and many other outstanding scientists. In following the artists and scientists of the Renaissance, all of these thinkers sought to create a new geometry of nature, intuitively understanding that, along with the well-known three-dimensional Euclidean coordinate space whose local basis is associated with the limits to a human being's clear field of vision, there are perspective and projective features of the world seen through the eyes of a person, whose visible event horizons are associated with the propagation speed of waves (including light), as well as Schwarzschild radii and Planck values.

From the point of view of mathematics, this new fractal (or fractional) concept of metrical determination turned out to be connected with the further development of infinite and infinitesimal aspects of integral and differential calculus during the transition to new visible event horizons determined by the propagation speed of waves (including light), Schwarzschild radii and Planck values.

O. V. Petrov, *The Earth's Free Oscillations*, https://doi.org/10.1007/978-3-030-67517-2

From the point of view of geometry, this single fractal (or fractional) concept of a three-dimensional Euclidean perspective and projective metrics was found to be associated with the three "great geometries":

- three-dimensional Euclidean geometry (third century BC) or spherical geometry with constant space curvature;
- infinite-dimensional Lobachevskian or hyperbolic geometry (1826) at speeds not exceeding the propagation speed of light with varying negative space curvature;
- infinite, but still finite-dimensional Riemann geometry (1854) with varying speeds reaching the speed of light propagation, the positive curvature of space in the direction of Schwarzschild gravitational radii and Planck quantities, objects on the surface of which the escape velocity reaches the speed of light propagation and they become black holes and dark matter of the universe invisible to an observer.

Recognising the innovative advantages of combining geometry with physics and solving the fundamental wave equation of nature when moving to new visible event horizons related to the speed of waves and light propagation, Schwarzschild radii and Planck scales, we are guided by Einstein's statement that the most correct description of space–time should be "… as simple as possible, but no simpler than this …" in presenting a space–time system of transformations of fundamental laws of nature in the framework of three "great geometries" in the form of three "great theorems":

Theorem 1 (Spatiotemporal transformations of Galileo) *Within the framework of three "great geometries" of the wave equation for natural vibrations of a string and a sphere with respect to special nodal and bifurcation reference points, this solution implies the same time frame during the transition from one inertial reference system to another. From the point of view of Newtonian physics, this comprises a system of equations:*

$$x' = x - vt$$

$$y' = y$$

$$z' = z$$

$$t' = t.$$

From a topological point of view, this is the solution of the famous Poincaré theorem "on the presence of at least two pairs of fixed points on an oscillating sphere" and the representation of the fractal dimension of oscillating spaces "… as reflections of points, lines and planes in themselves".

Theorem 2 (Spatiotemporal Lorentz transformation) *This solution is obtained within the framework of three "great geometries" of the wave equation for waves*

*propagating in a perspective measurement system with respect to perspective refer-
ence points, when the change in the negative and positive curvature of space cannot
exceed the propagation speed of light. From the point of view of physics, this consists
in the famous equation of Albert Einstein:*

$$x' = (x - vt)/\sqrt{1 - v^2/c^2}$$

$$y' = y$$

$$z' = z$$

$$t' = \left(t - vx/c^2\right)/\sqrt{1 - v^2/c^2}.$$

*"On the relationship of coordinates and time with the propagation speed of light"
and "on the discrete-wave or quantum nature of light". From a topological point of
view, this comprises a solution to the famous Poincaré "two-dimensional sphere"
theorem.*

Theorem 3 (Poincaré time–space transformations) *This solution obtained in the
framework of the three "great geometries" of the wave equation for standing waves
in a projective measurement system relative to negative and positive curved space
related to projective reference points, on the surface of which escape velocity is equal
to the speed of light propagation, thus comprising observer-invisible "black holes"
and the dark matter of the Universe. From the perspective of the general theory
of relativity, the surface of such bodies comprises a visible event horizon created
by discrete masses that are limiting with respect to the propagation speed of light,
whose dimensions correspond to the Schwarzschild radius in a projective observation
system, down to Planck scales and masses, along with the resonant interrelations
between them, which, decreasing inversely in proportion to the square of the distance
between such masses, determine the laws of universal gravitation. From the point
of view of topology, this represents the solution of the famous Poincaré theorems
"…on resonances" and "…three-dimensional spheres", as an everywhere-dense set
"without holes". Such sets are analogous to the perfect, everywhere-dense Cantor set
of points that fills the entire set segment, with the total mass of its points, nevertheless,
being equal to zero.*

The formulations and solutions of these theorems directly follow from the a
priori inherent contradictions between the three "great geometries", to which Henri
Poincaré was the first to pay attention. However, until now, no one has sufficiently
clarified the inextricable relationship with special nodal, bifurcational, perspective
and projective reference points, as well as the Galilean, Lorentzian and Poincaréan
spatiotemporal transformations of the fundamental laws of nature during the transition to new visible event horizons associated with the propagation speed of light at
Schwarzschild radii and Planck scales.

Using simple and illustrative examples describing the free oscillations of a string and the Earth, we were able to show that this inextricable relationship between the three "great geometries" and the three "great theorems" taking the form of the Galilean, Lorentzian and Poincaréan systems of spatiotemporal transformations of the fundamental laws of nature manifests itself with respect to special nodal, bifurcational, perspective and projective reference points, which at the same time serve as the origin of a unified system of three-dimensional Euclidean, perspective and projective metrics.

In the framework of the three "great geometries", it turned out that the very concept of three-dimensional space is defined in the form of a triaxial deformation ellipsoid involving presence of five types of special nodal and bifurcation points in Newtonian physics. Here, in contrast to the special nodal and bifurcation points of Newtonian dynamics, perspective and projective reference points possess affine properties of self-similar sets; that is, for them, there is no common reference point. These are bijections of one point to another, which form groups taking the form of fractal sets in perspective and projective measurement systems, related to the perception of the world around us through the speeds of wave and light propagation.

Thus, in solving wave equations in a perspective measurement system, when moving to a new event horizon related to the speed of light propagation, we will obtain five types of solutions for traveling waves or Lorentz space–time transformations. Solving the wave equation in the projective system of measurement, when moving to a new event horizon related to the propagation speed of light at Schwarzschild radii and Planck scales, we obtain five types of solution for standing waves, or spatiotemporal Poincaré transformations. All these formulations and solutions of the fundamental wave equation of nature make significant contributions both to the general and special theories of relativity, the currently popular "string theory", and, in general, to the unified theory of fractal measurement and spatiotemporal transformations of the fundamental laws of nature.

Within the framework of the unified theory of Galilean, Lorentz and Poincaré spatiotemporal transformations of the fundamental laws of nature with respect to special nodal and bifurcational, perspective and projective reference points and new visible event horizons related to the propagation speed of light involving Schwarzschild radii and Planck scales, all these fundamental laws of nature take on the dual form of structural-temporal laws, formulated in the same way as all physical laws since the great achievements of Maxwell's electromagnetic field theory, Einstein's general and special theories of relativity, as well as the laws of quantum mechanics. As Albert Einstein (1915) rightly remarked, all these dual laws "… link events that occurred here and now with events that occur a little later and in the immediate vicinity …" or, and most typically, at once and simultaneously. These relationships show that in the framework of the three "great geometries" that reflect the fractal (or fractional) structure of the space around us, the spatiotemporal transformations of the fundamental laws of nature taking the form of three "great theorems" of Galileo, Lorentz, and Poincaré simultaneously describe various aspects of the reality surrounding us.

We will not interpret this fractal (or fractional) dimension of space, or the network of duality empirically established in all areas of natural science, as a sign that, when creating a unified measurement system within the framework of the three "great geometries" of Euclid, Lobachevsky and Riemann and a unified theory of space–time transformations of fundamental laws nature in the form of the three "great theorems" of Galileo, Lorentz and Poincaré, we are on the right track. According to the figurative expression of the late Stephen Hawking, who for a long time occupied Lukasov's chair of mathematics at the University of Cambridge, inheriting this post from Isaac Newton and Paul Dirac, "… it would be the same as believing that God had placed fossil remains among the stones to confuse Darwin (and, at the same time, everyone else—author's note) on the question of the evolution of life."

Further Reading

Arnold VI (1990) Theory of catastrophes, 3rd edn, rev. M: Nauka, p 128

Bolt A (1984) Inside the Earth: evidence from earthquakes/M.: Mir, p 189. (in Russ.)

de Broglie L (1965) Revolution in physics: (New Physics and Quanta). In: de Broglie L, Trans. Baklanov SP, Kovrizhnykh LM, Polivanov MK (ed), 2nd edn. M: Atomizdat, p 231. (in Russ.)

Busse FH (1978) Non-linear properties of thermal convection. Res Prog Phys 41: 1931–1965

Darwin C (1859) On the origin of species by means if natural selection, or the preservation of favoured races in the struggle for life. London.

Deza MM, Deza E (2012) Encyclopedia of distances. Springer Science & Business Media, p 644. ISBN 9783642309588.

Einstein A (1923) [First published 1923, in English 1967]. Written at Gothenburg. Grundgedanken und Probleme der Relativitatstheorie. [Fundamental Ideas and Problems of the Theory of Relativity] (PDF) (Speech). Lecture delivered to the Nordic Assembly of Naturalists at Gothenburg, 11 July 1923. Nobel Lectures, Physics 1901–1921 (in German and English). Stockholm: Nobel-price.org. (published 3 February 2015)– via Nobel Media AB 2014

Einstein A (1936) Physik und Realitat. J Frankl Inst 221: 313–347. (Russian Translation: Einstein A (1967) Physics and reality. Einstein A (1967) Selection of scientific works in 4 volumes. M: Nauka, vol 4, p 200–227)

Ellis A (1999) Black holes–Part 1–History (Archive copy dated 6 October 2017 at Wayback Machine. Astron Soc Edinb J 39, лето.

van der Engdahl ER, Hilst RD, Buland RP (1998) Global teleseismic earthquake relocation with improved travel times and procedures for depth determination Bull. seism. Soc Am 88:722–743

Frank P (1909) Sitz. Ber. Akad. Wiss Wien 1909. II a, Bd 218. S373/esp, p 382

Ginzburg VL (1980) About physics and astrophysics. M Nauka, p 112. (in Russ.)

Gu WH, Morgenstern NR, Robertson PK (1994) Post-earthquake deformation analysis of wildlife site. J Geotech Eng 120(2):274–289

Haken H (1980) Synergetics. M: Mir, p 404. (in Russ.)

Haken H (1985) Synergetics. Hierarchies of instabilities in self-organising systems and devices. M. Mir, p 419. (in Russ.)

Hawking S (2007) The Universe in a nutshell. Trans A Sergeyev. Amphora , Saint Petersburg, p 218. (in Russ.)

Khain VE (1995) Core issue of modern geology (geology on the verge of XIX century). M: Nauka, p 190. (in Russ.)

Khain VE (2003) Core issues of modern geology, 2nd edn, revised. M: Nauchny Mir, p 348. (in Russ.)

Klimushkin DYu, Grablevsky SV (2001) Cosmology. M. (in Russ.)

Lamb H (1882) On the vibrations of an elastic sphere. Proc Lond Math Soc 13: 189

© The Editor(s) (if applicable) and The Author(s), under exclusive license to Springer Nature Switzerland AG 2021
O. V. Petrov, *The Earth's Free Oscillations*,
https://doi.org/10.1007/978-3-030-67517-2

Levin A (2005) History of black holes. Popular mechanics. OOO «Fashion Press». №11, p 52–62. (in Russ.)

Lobachevsky NN (1829) Kazan Bulletin. Part XXVI, books V and VI, May–June. (in Russ.)

Mandelbrot BB (1977) Fractals: form, chance and dimension. W.H. Freeman, San Francisco, p 273

Mandelbrot BB (1982) The fractal geometry of nature. W.H. Freman, San Francisco, p 460

Markov MA (2000) Selected works in 2 volumes. In: Matveev VA (ed), vol 1, M: Nauka, p 505, vol 2, M: Nauka, p 640. (In Russ.)

Petrov OV (1992a) Internal gravity waves of the Earth and non-linear paleogeodynamic dissipative structures. RAS report, vol 326, № 2, p 323–326. (in Russ.)

Petrov OV (1992b) Non-linear phenomena of thermo-gravitational instability and internal gravity waves of the Earth. RAS report, vol 326, № 3, p 506–509. (in Russ.)

Petrov OV (2007) The Earth's dissipative structures: fundamental wave properties of substance. VSEGEI Publishing, Saint Petersburg, p 304. (in Russ.)

Petrov OV (2019) The Earth's dissipative structures. Fundamental Wave Properties of Substance. © Springer Nature Switzerland AG.

Poincaré H (1974) Selected works in 3 volumes. M: Nauka. (in Russ.)

Poincaré H (1983) About science. In: Pontryagin LS (ed) M: Nauka, p 560. (in Russ.)

Poisson SD (1829) Memoire sur lequilibre et le movement des corps elastiques. Mem Paris Acad, 8.

Prigogine IR (2000) The end of certainty. Time, chaos and new laws of the nature. Scientific publishing centre «Regular and chaotic dynamics», Izhevsk, p 208. (in Russ.)

Prigogine IR, Stengers I (1986) Order out of chaos: man's new dialogue with nature. In: Arshinov VI, Klimontovich YuL, Sachkov YuV. M: Progress, p 432. (in Russ.)

Prigogine IR, Stengers I (2001) Time, chaos, quantum. To the solution of the time paradox. M: Editorial URSS, p 240. (in Russ.)

Riemann B (1948) Works. ML: GITTL, p 291. (in Russ.)

Schwarzschild K (1916) Uber das Gravitationsfeld eines Massenpunktes nach der Einsteinschen Theorie. Sitzungsberichte der Koniglich Preussischen Akademie der Wissenshaften 1, 189–196

Stokes GG (1847) On some cases of fluid motion. Philos Mag 31: 136–137

Turing AM (1952) F.R.S. The chemical basis of morphogenesis. Phil Trans R Soc Lond B 237(641): 37–72

Whitney H (1955) On singularities of mappings of Euclidean Spaces I. Mappings of the plane into the plane. Ann. Math 62: 374–410

Printed in the United States
by Baker & Taylor Publisher Services